MW01518281

RENEWABLE ELECTRICITY GENERATION

RESOURCES, STANDARDS, CHALLENGES

RENEWABLE ENERGY: RESEARCH, DEVELOPMENT AND POLICIES

Additional books in this series can be found on Nova's website under the Series tab.

Additional E-books in this series can be found on Nova's website under the e-books tab.

ENERGY SCIENCE, ENGINEERING AND TECHNOLOGY

Additional books in this series can be found on Nova's website under the Series tab.

Additional E-books in this series can be found on Nova's website under the e-books tab.

RENEWABLE ENERGY: RESEARCH, DEVELOPMENT AND POLICIES

RENEWABLE ELECTRICITY GENERATION

RESOURCES, STANDARDS, CHALLENGES

LEN G. IRWIN

AND

RAYMOND MACIAS

EDITORS

Nova Science Publishers, Inc.
New York

LIBRARY OF CONGRESS CATALOGING-IN-PUBLICATION DATA

Renewable electricity generation : resources, standards, challenges / editors, Len G. Irwin and Raymond Macias.
p. cm.
Includes bibliographical references and index.
ISBN 978-1-61942-678-8 (hardcover)
1. Renewable energy sources--United States. 2. Electric power production--United States. I. Irwin, Len G. II. Macias, Raymond.
TJ807.9.U6R43 2012
333.79'40973--dc23
2011050993

Published by Nova Science Publishers, Inc. † New York

CONTENTS

PREFACE

The United States faces important decisions about future energy supply and use. A key question is how renewable energy resources might be used to meet U.S. energy needs in general, and to meet U.S. electricity needs specifically. This book provides a summary of U.S. electricity generation potential from wind, solar, geothermal, hydroelectric, ocean-hydrokinetic, and biomass sources of renewable energy. An assessment of U.S. renewable electricity generation potential and how renewables might satisfy electric power sector demand is discussed, as are the challenges, issues and barriers that might limit renewable electricity generation deployment.

Chapter 1 - The U.S. energy sector is large and complex. Multiple energy sources, including fossil, nuclear, and several renewable sources, are used to produce energy products for multiple demand sectors (transportation, electricity, industrial, and residential/commercial). Today, fossil fuels are the dominant sources of energy, comprising 83% of total U.S. primary energy supply. Renewable energy sources, which can be used to generate electricity, produce liquid transportation fuels, and provide heating and cooling for industrial and residential/commercial sectors, provided 8% of total U.S. primary energy supply in 2009.

Chapter 2 - Federal lawmakers have recently considered several policies to alter the mix of fuels used to generate electricity in the United States. Those policies—referred to here as renewable or "clean" electricity standards— would lead to greater reliance on energy sources that produce few or no emissions of carbon dioxide (CO_2), the most prevalent greenhouse gas contributing to climate change. Renewable electricity standards would require a certain share of the nation's electricity generation (say, 25 percent) to come from renewable sources, such as wind or solar power. Clean electricity standards would require a certain percentage of the nation's electricity generation to come from renewable sources or from nonrenewable sources that reduce or eliminate CO_2 emissions, such as nuclear power, coal-fired plants that capture and store CO_2 emissions, and possibly natural-gas-fired plants. Many renewable and clean electricity standards already exist at the state level. This Congressional Budget Office (CBO) study—prepared at the request of the Chairman and Ranking Member of the Senate Committee on Energy and Natural Resources—examines how a federal renewable or clean electricity standard would change the mix of fuels used for electricity generation, the amount of CO_2 emissions, and the retail price of electricity in different parts of the United States. In particular, the study explores how some proposed features of such standards (such as various preferences, exemptions, and alternative compliance rules) would affect those outcomes and identifies underlying causes of uncertainty about such outcomes. The study also highlights key elements in designing a renewable or clean electricity standard that would help minimize its costs to U.S. households and businesses. In keeping with CBO's mandate to provide objective, impartial analysis, this report makes no recommendations.

Chapter 3 - This report responds to a July 2011 request to the U.S. Energy Information Administration (EIA) from Chairman Ralph M. Hall of the U.S. House of Representatives Committee on Science, Space, and Technology, for an analysis of the impacts of a Clean Energy Standard (CES). The request, as outlined in the letter included in Appendix A, sets out specific assumptions and scenarios for the study.

In: Renewable Electricity Generation   ISBN: 978-1-61942-678-8
Editors: L. G. Irwin and R. Macias   © 2012 Nova Science Publishers, Inc.

Chapter 1

U.S. RENEWABLE ELECTRICITY GENERATION: RESOURCES AND CHALLENGES[*]

Phillip Brown and Gene Whitney

SUMMARY

The United States faces important decisions about future energy supply and use. A key question is how renewable energy resources might be used to meet U.S. energy needs in general, and to meet U.S. electricity needs specifically. Renewable energy sources are typically used for three general types of applications: electricity generation, biofuels/bioproducts, and heating/cooling. Each application uses different technologies to convert renewable energy sources into usable products. The literature on renewable energy resources, conversion technologies for different applications, and economics is massive. This report focuses on electricity generation from renewable energy sources. In 2010, renewable sources of energy were used to produce almost 11% (7% from hydropower and 4% from other renewables) of the 4 million gigawatthours of electricity generated in the United States.

This report provides a summary of U.S. electricity generation potential from wind, solar, geothermal, hydroelectric, ocean-hydrokinetic, and biomass sources of renewable energy. The focus of this report is

---

[*] This is an edited, reformatted and augmented version of a Congressional Research Service publication, CRS Report for Congress R41954, from www.crs.gov, dated August 5, 2011.

twofold: (1) provide an assessment of U.S. renewable electricity generation potential and how renewables might satisfy electric power sector demand, and (2) discuss challenges, issues, and barriers that might limit renewable electricity generation deployment.

Data sources from 15 different organizations were reviewed to derive estimates of electricity generation potential. One key finding is that there exists no uniform national assessment of renewable electricity generation potential. No standard methods or set of assumptions are used to estimate renewable electricity generation potential. So even existing assessments for individual energy sources are difficult to compare objectively. In order to compare various estimates on an equivalent basis, CRS engaged experts in each renewable energy resource area to help normalize electricity generation potential estimates into a common metric: gigawatthours per year.

After surveying, researching, and normalizing all of the third-party electricity generation estimates, results indicate that renewable energy sources may, in principle, have the potential to satisfy a large portion of U.S. electricity demand. However, a number of potential barriers to large-scale deployment exist, including cost, power system integration, intermittency and variability, land requirements, transmission access, possible limits to the availability of key materials and resources, certain environmental impacts, specialized infrastructure requirements, and policy issues. Ultimately, the amount of renewable electricity generation in the U.S. may be dependent on the ability to address these deployment barriers. The Energy Information Administration projects that U.S. renewable electricity generation will increase from 11% today to between 14% and 15% in 2035.

As Congress considers policy options associated with increasing renewable electricity generation, policy makers may assess potential benefits such as emissions reduction, job creation, and global competitiveness, along with possible risks and consequences such as electricity cost and price increases, electricity delivery reliability, and environmental impacts associated with large-scale deployment of renewable electricity generation technologies.

INTRODUCTION

The U.S. energy sector is large and complex. Multiple energy sources, including fossil, nuclear, and several renewable sources, are used to produce energy products for multiple demand sectors (transportation, electricity, industrial, and residential/commercial). Today, fossil fuels are the dominant sources of energy, comprising 83% of total U.S. primary energy supply.

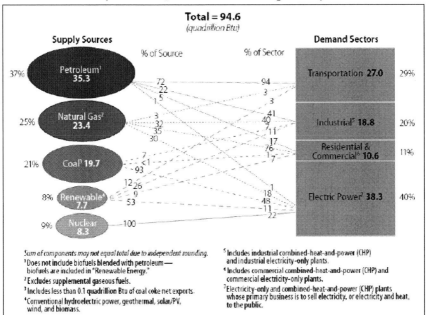

(Values are in Quadrillion Btu and Percentage of Total)

Source: CRS adaptation of Energy Information Administration, Annual Energy Review 2009, http://www.eia.doe.gov/totalenergy/data/annual/pdf/pecss_diagram_2009.pdf.

Figure 1. U.S. Primary Energy Flow by Supply Source and Demand Sector, 2009.

Renewable energy sources, which can be used to generate electricity, produce liquid transportation fuels, and provide heating and cooling for industrial and residential/commercial sectors, provided 8% of total U.S. primary energy supply in 2009 (see *Figure 1*).

The largest source of energy demand in the United States is the electric power sector, which consumed just over 40% of total U.S. energy supply in 2009. The U.S. electric power sector generates approximately 4 million gigawatthours of electricity each year. Like the total U.S. energy sector, electricity generation is dominated (89%) by fossil fuels and nuclear power.

Renewable electricity generation, including hydro, wind, solar, geothermal, and biomass, contributed 11% of total U.S. electric power in 2009 (*Figure 2*).[1] Most U.S. renewable generation comes from conventional hydropower, which has limited growth potential. Other renewable electricity sources constitute about 4% of U.S. generation, but have been growing more rapidly.

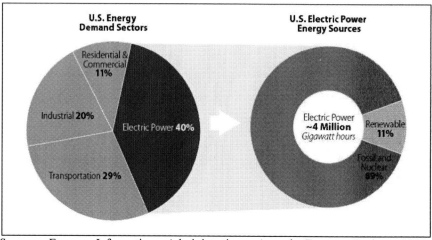

Figure 2. Supply Sources for U.S. Electric Power Sector.

The purpose of this report is to analyze the prospects, opportunities, and challenges for renewable energy sources to increase their contribution to the electric power sector.

There is growing interest in increasing the amount of renewable electricity generation to reduce the amount of fossil fuel consumption for U.S. electric power. That interest is driven by concerns about greenhouse gas emissions, the realization that economically recoverable fossil fuel supplies are ultimately finite, and the desire to position the United States as a global leader for renewable energy technology and manufacturing.[2] These concerns are counter-balanced by the fact that fossil fuel electricity generation has long been—and generally continues to be—the least expensive form of electricity generation, by the fact that the United States has access to considerable resources of coal and natural gas for electricity generation, and from the economic and cultural inertia of the existing infrastructure in place for coal and natural gas to be used in large quantities for electricity generation. Renewable electricity generation provides two advantages when compared to fossil generation: (1) it relies on energy sources that may not decline over time, and (2) it produces little or no net greenhouse gas emissions or other pollutants during use.[3] However, renewable electricity generation does have liabilities and implementation challenges that will be further discussed in this report.

This report addresses two fundamental questions about U.S. renewable electricity generation potential: (1) How much renewable electricity generation might be possible in the United States?[4] and (2) What technical, operational, and economic challenges might renewables encounter when considering large-scale deployment for electricity generation?

RENEWABLE ELECTRICITY CONCEPTS AND UNITS

Definition and Characteristics of Renewable Electricity

Renewable electricity is derived from renewable energy sources that "regenerate and can be sustained indefinitely."[5] This report does not use the term "clean energy" or "alternative energy," which are terms used by some to include renewable energy resources *plus* other sources that may emit little or no carbon dioxide during use, such as nuclear plants and coal-fired power plants equipped with carbon capture and sequestration capabilities. A discussion of biomass used to generate electricity is included, but biofuels are mentioned only briefly. This study is focused on wind, solar, hydroelectric, geothermal, biomass, and ocean/hydrokinetic energy sources used to generate electricity.

"Renewable energy" sources for electricity generation are often discussed as if they were a single entity, but renewable energy sources are more numerous and variable than fossil energy sources. Fossil fuels comprise oil, natural gas, and coal. The three major types of fossil fuels are extracted from the earth's crust by drilling or mining. Each of these fuels has very high energy density and is used primarily through combustion to exploit the heat produced. Renewable energy sources are more numerous and diverse and, thus, harnessing renewable energy requires a number of different technologies. Some of the distinctive characteristics of renewable energy are:

- *Renewable energy sources for electricity generation are numerous.* Sun, wind, flowing water in streams, flowing water in tidal channels, wave action in oceans, the earth's natural heat, biological materials, and others comprise the current portfolio of renewable energy sources, and additional renewable sources may be identified in the future.
- Each renewable energy source may be exploited in multiple ways to generate electricity using different technologies and materials. For example, the energy of the sun may be used by concentrating the

energy to generate steam that drives electric turbines (concentrating solar power), or the energy of the sun may be converted directly to electricity using semiconducting materials (photovoltaics). Furthermore, photovoltaic electricity may be produced using solar panels that consist of crystalline silicon, cadmium telluride, or other materials, and each material has unique characteristics.

- Renewable energy sources for electricity generation are naturally dispersed with relatively low energy densities. Fossil energy sources are typically concentrated as liquids or solids by millions of years of natural heating and pressure processes, which result in relatively high energy density that is accessible in wells or mines. In contrast, renewable energy sources are typically diffuse and require multiple technologies and management systems to gather and concentrate the resources.

- Each renewable electricity generation project/installation can vary in size. Renewable electricity generation systems are being installed in large, megawatt-scale projects that feed electricity into the electric grid for consumption along with electricity from other sources. In fact, the largest electric power plant in the United States is a hydroelectric facility, Grand Coulee Dam, which has a capacity of 7.08 GW.[6] At the same time, individual homes are being powered by small, kilowatt-scale rooftop solar panels. Also, wind turbines may be large, up to 5 megawatt (MW) utility-scale turbines, or small, approximately 5 kilowatt (kW) residential scale units.

Renewable Electricity Terminology and Units

This section defines terms and units used to describe and quantify renewable energy sources, and electricity generation potential from these sources, and how renewable energy might be compared to other forms of energy. Although this report focuses on renewable energy, discussions of fossil fuel units and consumption are included to facilitate comparisons with renewable forms of energy.

Measuring Energy: Fossil versus Renewable

Fossil fuels have traditionally been measured and marketed in the units of the physical material— barrels (42 gallons) of oil, short tons (2,000 pounds) of coal, or cubic feet of natural gas— transported to the point of end use. The use of volume or weight for measuring fossil fuels makes it challenging to compare the energy content among fossil fuels, and also contributes to the difficulty in clearly communicating the amounts of renewable energy that will be needed to replace fossil fuels. Each fossil fuel unit of measure has a corresponding energy content, which is typically expressed in terms of British Thermal Units (Btu).[7]

With the exception of biomass (typically measured in tons), each renewable energy source has its own unit of measure that may not be expressed as volume or weight. For example, wind energy is typically expressed in terms of *wind speed* (reported as meters per second); solar energy is typically expressed in terms of daily *insolation* (reported as kilowatthours per meter-square per day); hydroelectric is derived from flowing water, typically expressed in terms of water *flow rate* and *velocity*. In order to estimate annual electricity generation potential from renewable energy sources, experts must make assumptions about conversion equipment efficiencies and annual hours of operation.

Expressing Renewable Electricity Generation Potential: Watthour

Renewable electricity generation potential is typically expressed in terms of watthours (see text box below). A watthour (Wh) is a unit of electrical energy that can be generated, distributed, and consumed. A watthour can also be purchased and/or sold. For example, a residential electricity bill is typically calculated by multiplying the number of kilowatthours (kWh) consumed by a residence times the rate per kilowatthour charged by the electric power provider.[8] In 2009, U.S. total electricity net generation was approximately 4 million gigawatthours.[9] For the purpose of this report, renewable electricity generation potential, for all renewable energy sources, is expressed in terms of annual gigawatthours (GWh).

Authoritative Data Sources for Renewable Energy Resources

Various renewable electricity resource estimates for the United States are calculated by different institutions that use different processes, methodologies, and assumptions. No uniform methodologies exist for estimating and comparing the resource potential of different forms of renewable energy that might be used to generate electricity (see text box).

**Power Versus Energy: What's the Difference between
a Watt and a Watthour?**

Some energy reports provide statistics in units of power while other reports use units of energy. Power and energy are related, but they are not the same thing. Energy equals power multiplied by the amount of time the power is applied. Conversely, power is the rate at which energy is produced or consumed. Power is measured in watts, energy is measured in watthours. An electrical generator with 50 megawatts of power (or nameplate capacity) would generate 50 megawatthours of electrical energy for each hour it operates. The power capacity of a generator conveys only the size of the device, thus, when it's not operating, a generator does not produce any energy even though the power capacity remains the same. Power capacity is a critical variable when selecting a device to do a specific job, but the energy produced by the device depends on the amount of time it operates. The report examines total energy production with little regard to the size of the devices that produce it.

Traditional fossil fuel energy resource assessments are conducted through detailed geologic studies and the application of rigorously vetted methodologies. In contrast, most renewable electricity generation resource estimates are subject to the unique methods and assumptions of the organization conducting the assessment. Fossil energy resource estimates typically classify resources into categories such as: resource base, technically recoverable, economically recoverable, and reserves.[10] In principle, renewable energy resources should be measurable using similar analysis of the natural processes (wind, solar insolation, water flow, geothermal heat, etc.), adjusted for the effectiveness of the respective energy extraction technologies, and then couched in economic terms based on economic conditions and parameters.

In reality, it is very difficult, time consuming, and expensive to collect high quality data for wind, solar, stream flow, geothermal and biomass energy at a fine scale over the entire nation on an hourly, daily, or seasonal basis, as appropriate. In addition, the basic physics are different for extracting energy from solar, wind, hydro, geothermal, and biomass sources. Those physical differences give rise to different technologies, and many renewable energy technologies exhibit dramatically different performance according to geographic location and time of day or time of year. Therefore, estimating the amount of each type of renewable energy that is available to the nation is a challenging task. Examination of the literature reveals that estimates of available renewable energy resources vary widely. Attempting to compare estimates for different types of renewable energy multiplies those challenges.

The most reliable data for the various renewable energy sources come from national data collection programs from federal agencies. For example, the National Renewable Energy Laboratory, funded by the Department of Energy and its partners, operates programs designed to collect such data and has worked to identify the areas within the United States that are optimally suited for exploitation of various renewable energy sources. Other federal and state agencies, federal labs, and academic institutions also collect, analyze, and report renewable energy resource data. These data and estimates change over time as data collection technologies advance and understanding of the natural processes improves. Nevertheless, comparing renewable energy assessments from different sources is difficult, and a complete and comprehensive assessment of all available renewable energy resources for the nation does not yet exist. Collection of high quality data on renewable energy sources at a fine scale over broad ranges of time and geography will likely be an ongoing need for the nation. *Table 1* summarizes sources of information reviewed for this report.

Table 1. U.S. Renewable Electricity Generation Potential—Information Sources

Renewable Electricity Resource	Sources of Data Reviewed
Wind	National Renewable Energy Laboratory (NREL); American Wind Energy Association (AWEA); Department of Energy (DOE) Office of Energy Efficiency and Renewable Energy (EERE).
Solar	NREL; Solar Energy Industries Association (SEIA); DOE EERE.
Hydro	DOE EERE, Oak Ridge National Laboratory (ORNL); Idaho National Laboratory (INL); National Hydropower Association (NHA).
Geothermal	United States Geological Survey (USGS); NREL; Massachusetts Institute of Technology (MIT); Geothermal Energy Association (GEA).
Ocean-Hydrokinetic	Electric Power Research Institute (EPRI); DOE EERE; New York University; Ocean Renewable Energy Coalition (OREC).
Biomass	DOE EERE; United States Department of Agriculture (USDA); NREL, Biomass Power Association (BPA).

Source: CRS.

**How Much Renewable Energy Is Available? It Depends ...
and It Can Change**

The overriding goal of this report is to provide Congress with accurate, comparable, and current U.S. renewable electricity resource estimates using currently available data. Answering the question "How much renewable electricity is possible in the United States?" is the primary objective. However, the answer to this key question is "it depends, and it will very likely change over time."

No centralized authoritative body or organization currently exists to develop and enforce standards for renewable electricity resource assessment methodologies and assumptions used to calculate estimates. Renewable electricity resource estimates come from multiple organizations. As a result, renewable electricity generation potential estimates are derived using different methodologies and different assumptions which, in turn, produce different estimates.

Estimates for renewable electricity generation potential in the United States depend on several factors such as the methodology used to calculate the estimates and certain assumptions that can have a major impact on the calculation results. With regard to methodologies used, resource estimates surveyed for this report came from several different organizations that include federal labs, industry organizations, and academic institutions. Each organization typically uses a unique methodology to calculate resource estimates. Therefore, comparing all of these estimates on an "apples-to-apples" equivalent basis is a challenge.

Key assumptions made for calculating renewable electricity generation potential can also have a major impact on resulting estimates. Geothermal electricity is a good example of how assumptions can impact renewable electricity generation estimates. Both USGS and MIT have published reports that estimate the amount of electricity generation potential from U.S. geothermal resources.[11] However, MIT estimates are more than 10 times larger than those from USGS. Two key assumptions explain most of this discrepancy: (1) Resource depth: The USGS geothermal study only considered geothermal potential at depths of 6 kilometers below the earth's surface whereas the MIT report considered depths of 10 kilometers, and (2) Which U.S. states were included: The USGS study only included 14 western states, Alaska, and Hawaii, while the MIT study included all 50 states. Thus, understanding assumptions for understanding and comparing the various resource potential estimates is critical.

Understanding certain exclusions for the various resource potential estimates is also important. Many of the studies surveyed for this report excluded certain areas from development based on several factors (national parks, urban areas, etc.). However, the types of exclusions and the constraints that result from exclusions vary. Comparing wind estimates and hydroelectricity estimates is one example. Wind electricity generation potential estimates exclude certain land areas. After these exclusions are taken into account, the NREL study referenced for this report assumes that wind projects can be built anywhere as long as the wind resource is large enough to meet certain electricity production levels. Hydroelectricity generation estimates, on the other hand, also include certain land area exclusions but apply additional filters such as the location being within one mile of a road and a transmission line. If identical exclusions were applied to all renewable electricity generation resource assessments, resource potential results may be quite different.

Further, estimates for renewable electricity generation potential in the U.S. will likely change over time as resource estimate methodologies improve, renewable electricity generation technologies are developed and commercialized, and better information about the magnitude and quality of renewable energy resources is made available. As a result, estimates of renewable electricity generation potential could either go up or down in the future.[12]

U.S. RENEWABLE ELECTRICITY USE AND POTENTIAL

This section provides a brief overview of current U.S. renewable electricity generation, followed by a series of discussions of specific renewable electricity technologies. Current electricity generation, estimated potential generation, and deployment challenges are discussed for wind, solar, geothermal, hydroelectric (hydro), ocean-hydrokinetic, and biomass energy sources.

Summary of Current U.S. Renewable Electricity

In 2009 renewable energy resources provided 11% of U.S. electricity net generation. Renewable electricity was derived from wind, solar, geothermal,

(Percentage of each renewable source)

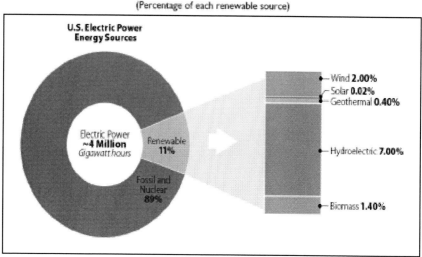

Source: Energy Information Administration, Annual Energy Review 2009, http://www.eia.doe.gov/totalenergy/ data/annual/pdf/pecss_diagram_2009.pdf.

Notes: Renewable electric power percentages may not add to 11% because of independent rounding error.

Figure 3. U.S. Electricity Generation from Various Renewable Sources, 2009.

hydro, and biomass energy sources. The largest source of renewable electricity was hydro. Wind and biomass each contributed between 1% and 2% of total U.S. electricity net generation. Solar and geothermal electricity generation contributed relatively small amounts to the renewable electricity portfolio mix (*Figure 3*).

Future Renewable Electricity Generation Potential

The following sections discuss the estimated range of electricity generation potential from wind, solar, geothermal, hydro, ocean-hydrokinetic, and biomass renewable energy sources. A discussion of technology and cost considerations is presented for each respective renewable source of electricity. As discussed above, comparing renewable electricity generation resource estimates is a challenging task. The approach used to derive the resource estimate range for each technology is described in the footnotes to each renewable energy source section. Table 2 provides a summary of U.S. renewable electricity generation potential, based on the research and analysis performed for this report, current and projected renewable electricity generation potential, cost of electricity estimates, and a summary of key challenges for each renewable energy source. As the table shows, renewable electricity generation potential is compared to 2009 total U.S. net generation of approximately 4 million gigawatthours (GWh). This approach was used in order to indicate the maximum electricity generation contribution that might be available from each renewable energy source. Furthermore, the reader is advised that levelized cost of energy (LCOE) estimates presented in *Table 2* only reflect electricity costs associated with capacity additions in EIA's Annual Energy Outlook 2011. For more information, see the "Levelized Cost of Energy (LCOE)" section below.

Research and analysis conducted for this report indicates that renewable energy sources may, theoretically, have the potential to satisfy a large portion of U.S. electric power needs. However, numerous technical, operational, economic, and practical challenges will likely be encountered, which may ultimately limit the potential contribution of renewable electricity generation. These challenges are discussed in a following section. Furthermore, while the potential for renewable electricity generation in the country is vast, EIA Annual Energy Outlook 2011 reference case projections indicate that renewables will contribute between 14% and 15% of total U.S. electricity generation by 2035.[13] Also, the quality of resources estimates is different for each renewable technology, and these estimates may change as new data are collected and new assessments are conducted. The current estimates represent a snapshot in time and must be continually updated as additional data become available.

Wind

U.S. Resource Estimates

U.S. wind energy resource estimates are highly dependent on certain assumptions used to calculate them, and users of those estimates should pay careful attention to the underlying assumptions. Turbine height and capacity factor assumptions can have major impacts on wind resource estimates. For example, winds are generally stronger at greater heights above the ground. As a result, wind resource estimates at 100 meters are likely to be greater than those at 50 meters.

Table 2. Summary of U.S. Renewable Electricity Resources and Challenges

Electricity Generation Potential	Wind[a]		Solar[b]		Geothermal		Hydro		Ocean–Hydrokinetic		Biomass[c]	
	Low	High	Low	High	Low	High	Low	High	Low	High	Low	High
Electricity Potential (GWh/yr) * Total estimated resource potential	32,500,000	61,400,000	4,000,000	56,300,000	927,791	36,991,864	558,145	613,333	287,850	2,161,350	125,730	1,428,780
% of 2009 total U.S. generation	>100%	>100%	100%	>100%	23%	>100%	14%	16%	7%	55%	3%	36%
Current and Forecasted Generation												
2009 Generation (GWh)	73,886		891		15,009		273,445		not available		54,493	
% of 2009 total U.S. generation	1.87%		0.02%		0.38%		6.92%		not available		1.38%	
EIA LCOE[d] $/MWh ** Only for capacity additions forecasted in AEO 2011	$82 to $349		$159 to $642		$92 to $116		$59 to $121		not available		$99 to $134	
2035 Generation (GWh) *** EIA AEO 2011 forecast (reference case)	160,880		3,970		49,190		310,590		not available		47,440	
% of 2035 total est. generation	3.48%		0.09%		1.06%		6.70%		not available		1.02%	
Deployment Challenges, Issues, and Barriers?												
Power System Integration	Yes		Yes		No		No		Yes		No	
Transmission	Yes		Yes		Yes		Yes		Yes		Yes	
Cost	No/Yes		Yes		Yes		No/Yes/		Yes		Yes	
	- Onshore wind costs		- Currently the highest		- Enhanced geothermal		-One of the oldest and		- Actual cost of		- Cost of electricity	

Electricity Generation	Wind[a]		Solar[b]		Geothermal		Hydro		Ocean-Hydrokinetic		Biomass[c]	
	Low	High	Low	High	Low	High	Low	High	Low	High	Low	High
Potential	are in the range of fossil electricity costs - Offshore costs are higher		cost source of renewable electricity		system (EGS) costs are estimates only; NREL indicates EGS LCOE could be as high as $1,000/MWh		lowest-cost sources of renewable electricity - Emerging small/low-head hydro costs unknown		ocean and hydrokinetic electricity is unknown		can be impacted by logistics and feedstock quality	
Intermittency/ Variability	Yes - Wind resources can vary on an hourly, daily, and/or annual basis		Yes - Cloud coverage and other weather events can degrade solar technology performance; solar energy not available at night		No - Geothermal electricity production can be predictable and may operate at high capacity factors		Yes - Hydropower resource can vary based on annual rain/snow fall		Yes/No - Wave energy resources can vary based on the amount of wind; tidal energy may be predictable		No - Biomass electricity plants can operate at high capacity factors and may provide baseload power	
Technology	No/Yes - Onshore wind technology considered commercial - Offshore wind may require further technology development to operate in harsh ocean environment		Yes - Emerging PV technologies that may improve efficiencies & reduce costs - Some CSP technologies may require further engineering, development, and demonstration before being commercial		Yes - EGS, the largest potential source of geothermal electricity, technology is not yet commercially available - Specialized drilling equipment may be required		Yes - Small and low-head/lowpower technologies are being developed and matured but some are not yet commercially available		Yes - Ocean and hydrokinetic energy technologies are considered "emerging" with no commercially available electricity generation technologies		No/Yes - Biomass combustion technology might be considered commercial - Some technical issues (tar production, equipment fouling etc.) may have to be addressed	

Table 2. (Continued)

Electricity Generation Potential	Wind[a]		Solar[b]		Geothermal		Hydro		Ocean-Hydrokinetic		Biomass[c]	
	Low	High	Low	High	Low	High	Low	High	Low	High	Low	High
Environmental Impact	Yes - Land use, habitat and scenic disturbance, noise, and bird mortality are potential environmental issues associated with wind projects		Yes - Water use requirements for some solar thermal technologies - Land use and associated habitat disturbance - Mobilization of trace metals		Yes - Water use; discharge of metals and toxic gas - Ground/surface water pollution - Land subsidence and seismicity		Yes - Ecosystem changes; fish migration and mortality - Habitat damage; water quality degradation		Yes - Alteration of currents and waves; alteration of sediment disposition: habitat impacts; noise; electromagnetic fields; toxicity of lubricants and other fluids; animal injury from moving parts; degradation of water quality		Yes- Biomass combustion for electricity generation emits NOx, CO_2, and other emissions - Land use/change associated with biomass production - Carbon neutrality of biomass combustion is a possible issue	
Infrastructure	Yes - Offshore wind may require specialized vessels, portside infrastructure, under-sea transmission, etc.		Unknown - Information regarding potential infrastructure issues not available		Unknown -Information regarding potential infrastructure issues not available		Unknown - Information regarding potential infrastructure issues not available		Yes - Specialized infrastructure may be needed to install and maintain operational projects		Yes - Logistics infrastructure may be needed to gather and process biomass material	
Materials and Resources	Yes - High volumes of steel, concrete, and rare-earth metals may be needed to support large scale wind deployment		Yes - High volumes of steel and concrete may be needed to support large-scale deployment of utility -scale solar; silicon, tellurium, cadmium,		Unknown - Information regarding potential materials and resources issues not available		Unknown - Information regarding potential materials and resources issues not available		Yes - Materials that can operate for long periods of time in a corrosive ocean environment may need to be developed		Yes - Economical electricity generation from biomass may require adequate biomass resources within a defined geographic area	

Electricity Generation	Wind[a]		Solar[b]		Geothermal		Hydro		Ocean-Hydrokinetic		Biomass[c]	
Potential	Low	High	Low	High	Low	High	Low	High	Low	High	Low	High
			silver, and other commodities may be required for large-scale PV deployment									

Source: CRS; Various sources as identified and referenced in the respective sections of this report.

[a] Includes both onshore and offshore wind.

[b] Includes both photovoltaic and concentrating solar.

[c] Does not include liquid biofuels used for transportation.

[d] LCOE = Levelized Cost of Energy.

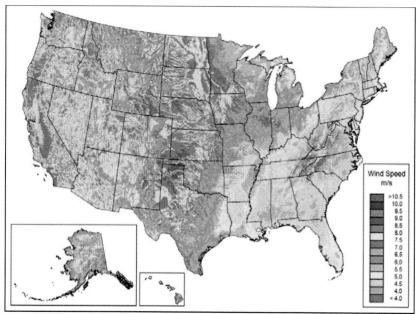

Source: http://www.windpoweringamerica.gov/wind_maps.asp.

Figure 4. U.S. Onshore Wind Energy Resources, 80 Meter Turbine Height.

Wind energy resources in the United States are typically categorized as either "onshore" or "offshore." According to National Renewable Energy Laboratory (NREL) estimates, onshore wind electricity generation potential for the 48 contiguous United States ranges from 22.5 million gigawatthours to 46.9 million gigawatthours annually.[14] Wind resources can vary state by state and region by region. Based on NREL estimates, the largest onshore U.S. wind energy resources are located in the middle of the country (see *Figure 4*).

NREL has also calculated estimates for U.S. offshore wind energy resources.[15] Based on NREL estimates, offshore wind energy resource potential may range between 10 million GWh and 14.5 million GWh annually.[16] *Figure 5* illustrates how offshore wind energy resources vary by location. However, in its February 2011 National Offshore Wind Strategy report, the DOE EERE states that "the offshore wind resource is not well characterized."[17] This uncertainty indicates that additional work may be needed to more accurately assess U.S. offshore wind potential.

(Excluding Alaska)

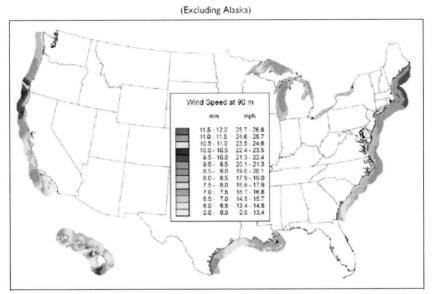

Source: http://www.windpoweringamerica.gov/windmaps/offshore.asp.

Figure 5. U.S. Offshore Wind Energy Resources, 90 Meter Turbine Height.

Technology and Cost Considerations

Onshore wind energy conversion technology is generally considered to be commercially available[18] and many projects are able to attract debt and equity investment capital for project development. General Electric and Siemens were the top two manufacturers of wind turbines installed in the U.S. during 2010.[19] Typical wind turbines have a rated capacity between 1 megawatt and 3 megawatts, and the general industry trend is to continue increasing the size and capacity of individual wind turbines in order to operate at greater heights (taller towers) and realize economies of scale by generating more watthours from a single unit. Offshore wind energy technology faces some technical challenges associated with operating in a corrosive marine environment and installation of equipment at various water depths. The Department of Energy (DOE) operates a Wind Power Program aimed at wind research and development needs.[20]

The cost of wind-generated electricity can vary based on a number of technology, performance, operational, and financial factors. These factors are discussed in the "Levelized Cost of Energy" section below. Assumptions made for these factors can result in significant differences among cost-of-electricity estimates. In its Annual Energy Outlook 2011, EIA estimates onshore wind electricity costs to range from $82-$115 per megawatthour (MWh) and offshore electricity costs between $187-$349 per MWh.[21] *Figure 12* provides a comparison of costs for conventional (fossil and nuclear) and renewable electricity generation.

Solar

U.S. Resource Estimates

Every U.S. locale receives sunlight during a calendar year, of course, but the amount of radiation that reaches a given point at a particular time can vary based on factors that might include geography (including latitude), time of day, season, landscape, and weather.[22] Two authoritative sources for U.S. solar resources are the National Solar Radiation Database (NSRDB), and NREL and State University of New York at Albany (SUNYA) satellite-derived solar resources.[23]

Quantifying solar resource data, in terms of annual electricity generation potential, is complicated by several factors.[24] First, two different methods are used for capturing and converting solar energy into electricity: (1) concentrating solar power (CSP), and (2) photovoltaic (PV) solar power.[25] Second, solar radiation has different components that may be better suited for different collector types.[26]

Third, different system configurations are used to collect data and calculate solar resource estimates.[27] Finally, different types of CSP and PV technologies, with different cost, efficiency, and performance characteristics, are available for solar energy conversion.[28]

CSP resources vary throughout the country, with most of the highest quality resource located in the southwestern United States (see *Figure 6*). CSP electricity generation is typically better suited for large-scale (greater than 10 MW) power generation projects. NREL and Sandia National Laboratories estimate that U.S. CSP electricity generation potential is approximately 16.3 million GWh.[29]

This estimate is the result of applying a set of filters to existing CSP resource data in order to calculate CSP electricity generation potential.[30] NREL and Sandia CSP estimates include electricity generation potential in seven U.S. states.[31]

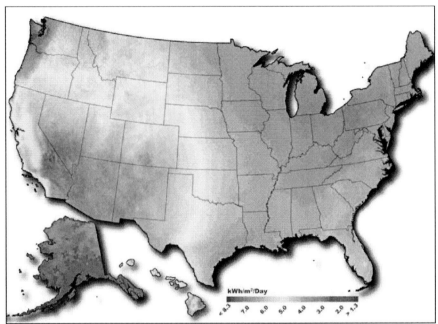

Source: National Renewable Energy Laboratory (NREL), available at http://www.nrel.gov/gis/images/ map_csp_national_lo-res.jpg.
Notes: Annual average direct normal solar resource data are shown. The data for Hawaii and the 48 contiguous states are 10km satellite modeled dataset (SUNY/NREL, 2007) representing data from 1998-2005. The data for Alaska are a 40 km dataset produced by the Climatological Solar Radiation Model (NREL, 2003); kWh/m2/Day = kilowatthour per square meter per day.

Figure 6. U.S. Concentrating Solar Resource.

Photovoltaic resources also vary throughout the country and, much like CSP, the highest quality PV resources are located in the southwestern United States (see *Figure 7*). PV systems offer flexibility in terms of project size and can be used for residential, commercial, and utility-scale applications. As of July 2011, estimates of the technically available and economically recoverable solar photovoltaic resource, in terms of annual GWh, were not available. However, previous work by NREL provides some indication about solar generation potential. NREL researchers analyzed land-use requirements for generating 100% of U.S. electricity and estimated that 0.6% of total U.S. land area would be needed to satisfy current demand load using a "base system configuration."[32] NREL has also estimated land-use requirements for a variety of other system configurations.[33] Calculating total PV generation based on this

NREL analysis is somewhat complicated, but it may be reasonable to assume that solar PV could theoretically generate 10 times the amount of current U.S. demand, although realizing this amount of electricity generation may be limited by several factors, particularly cost and power system integration.[34] Extrapolating from the NREL analysis, U.S. annual solar PV generation potential may be equal to approximately 40 million GWh. NREL has also evaluated the generation potential of residential and commercial rooftop PV systems and estimates that, under a base-case scenario, approximately 819,000 GWh of electricity could be generated each year using existing rooftop space.[35]

Technology and Cost Considerations

The most commonly used CSP technology in the United States is the parabolic trough. Of the 509 megawatts of U.S. installed CSP capacity, approximately 98% uses parabolic trough technology.[36] Crystalline silicon is the most commonly used PV technology.[37] While some CSP and PV technologies might be considered commercially available, there are a number of research and development activities within CSP and PV markets.[38] Generally speaking, most CSP and PV R&D work is focused on improving system-level efficiencies and reducing system costs. Storage, demand response, and other "smart-grid" technologies may further enable large-scale solar deployment.[39]

The cost of solar electricity has been a challenge faced by both CSP and PV technologies. Solar electricity, according to the Energy Information Administration, is the highest-cost source of electricity generation, with CSP costs ranging from $192-$642 per MWh and PV costs ranging from $159-$324 per MWh.[40] DOE is funding an initiative, known as the SunShot program, which aims to reduce the cost of PV electricity generation to $60 per MWh.[41] *Figure 12* provides a comparison of costs for conventional (fossil and nuclear) and renewable electricity generation.

Geothermal

U.S. Resource Estimates

Geothermal energy is present throughout the entire country, with most of the highest-quality geothermal resources generally located in the western United States, Alaska, and Hawaii.[42]

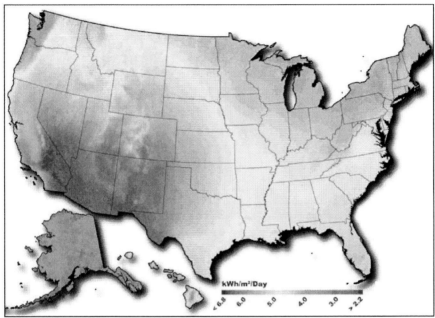

Source: National   Renewable   Energy   Laboratory   (NREL),   available   at
      http://www.nrel.gov/gis/images/ map_pv_national_lo-res.jpg.
Notes: Annual average solar resource data are shown for a tilt-latitude collector. The
      data for Hawaii and the 48 contiguous states are a 10 km satellite modeled dataset
      (SUNY/NREL, 2007) representing data from 1998- 2005. The data for Alaska are
      a 40 km dataset produced by the Climatological Solar Radiation Model (NREL,
      2003); kWh/m2/Day = kilowatthour per square meter per day.

Figure 7. U.S. Photovoltaic Solar Resource.

However, all states may have geothermal electricity generation potential
through the use of enhanced, or engineered, geothermal systems (EGS)
technology.[43] USGS, NREL, and the Massachusetts Institute of Technology
(MIT) each have published estimates for U.S. geothermal electricity
generation. Generally, geothermal electricity generation resources are
classified into three categories: (1) identified resources, (2) undiscovered
resources, and (3) enhanced geothermal systems.[44] *Table 3* provides a
summary of USGS, NREL, and MIT potential geothermal capacity estimates
for these resource types along with annual electricity generation potential in
GWh.

Table 3. U.S. Geothermal Electricity Generation Potential

	Identified Resource		Undiscovered Resource		Enhanced Geothermal Systems	
	Low	High	Low	High	Low	High
USGS[a] Capacity (MW-e) Electricity Generation[b] (GWh/yr)	3,675 29,618	16,457 132,630	7,917 63,805	73,286 590,627	345,100 834,369	727,900 1,759,888
NREL[c] Capacity (MW-e)	n/a	6,390	n/a	30,030	n/a	15,000,913
Electricity Generation[b] (GWh/yr)	n/a	51,498	n/a	242,018	n/a	36,268,607
MIT[d]						
Capacity (MW-e)	n/a	n/a	n/a	n/a	1,249,000	12,486,000
Electricity Generation[b] (GWh/yr)	n/a	n/a	n/a	n/a	3,019,782	30,188,151

Source: U.S. Geological Survey, National Renewable Energy Laboratory, MIT. See
specific references below.

Notes: Electricity generation potential estimates represent what might be technically
recoverable and do not include any filters for economic factors. "Identified" and
"Undiscovered" resources generally represent conventional geothermal resources
where naturally occurring conditions (high temperature and permeability) allow
for extraction of geothermal energy. "Enhanced Geothermal Systems" require
engineering of rock permeability to create geothermal energy extraction
conditions. Electricity generation numbers were calculated by CRS using a 92%
capacity factor for each geothermal capacity estimate. USGS low estimates for
each resource category represent resources that have a 95% probability of being
available. USGS high estimates for each resource category represent resources
that have a 5% probability of being available. NREL analysis provided a single
number for identified, undiscovered, and EGS resource estimates, respectively.
The MIT study focused on the potential of EGS in the United States. The large
difference between MIT's low and high estimates reflect an assumption made for
the EGS energy recovery factor (2% for the low estimate, 20% for the high
estimate).

[a] "Assessment of Moderate- and High-Temperature Geothermal Resources of the
United States," U.S. Geological Survey, 2008, available at http://pubs.usgs.gov/
fs/2008/3082/pdf/fs2008-3082.pdf.

[b] Annual electricity generation potential assumes that all potential geothermal
resources are developed and operating. This is highly unlikely since EGS systems
may result in resource depletion over a 30-40 year operating life; regeneration of
this resource is estimated to take approximately 100 years. As a result, EGS
estimates for GWh/yr were discounted by a factor of 0.3 in order to calculate
sustainable electricity generation potential based on a 30-year depletion and 100-
year regeneration profile.

[c]. Augustine, K. Young, and A. Anderson, "Updated U.S. Geothermal Supply Curve," National Renewable Energy Laboratory, Conference Paper presented at Stanford Geothermal Workshop, February 1, 2010, available at http://www.nrel.gov/docs/ fy07osti/41073.pdf.

[d] "The Future of Geothermal Energy: Impact of Enhanced Geothermal Systems (EGS) on the United States in the 21st Century," Massachusetts Institute of Technology, 2006, available at http://geothermal.inel.gov/ publications/future_of_geothermal_ energy.pdf.

MW-e = megawatt electrical generating capacity.

GWh/yr = gigawatthours per year.

USGS, NREL, and MIT each have a different estimate for U.S. geothermal electricity generation potential, especially with regard to enhanced geothermal systems. Two primary factors account for the differences in estimates: (1) USGS estimates are confined to western U.S. states, Hawaii, and Alaska, while NREL and MIT estimates include potential electricity generation from all 50 states, and (2) USGS estimates are for resource depths between 3 kilometers (km) and 6 km, while NREL and MIT estimates are for resource depths between 3 km and 10 km.[45] This disparity in resource estimates illustrates how assumptions can significantly alter the assessment results.

Figure 8 shows conventional geothermal sites and the estimated relative suitability of EGS geothermal energy recovery throughout the U.S.

Technology and Cost Considerations

For conventional hydrothermal geothermal resources, four commercially available technologies are available for generating electricity: (1) flash power plants, (2) dry steam power plants, (3) binary power plants, and (4) flash/binary combined cycle.[46] As of April 2011, U.S. geothermal installed capacity was 3,102 MW, which represents approximately 0.3% of total U.S. electricity capacity. In 2009, 15,009 GWh of electricity was generated from geothermal energy sources. The majority of existing geothermal capacity is located in California.[47] Enhanced Geothermal Systems (EGS) technology could potentially enable large-scale deployment of economically recoverable geothermal electricity generation.[48] However, EGS technology has not been demonstrated at scale and is not yet commercially available.[49]

Several factors can influence the cost of geothermal electricity. These factors include the resource quality (temperature and volume), resource depth, drilling costs, and geothermal equipment costs.

(Locations of identified hydrothermal sites and favorability of deep enhanced geothermal systems [EGS])

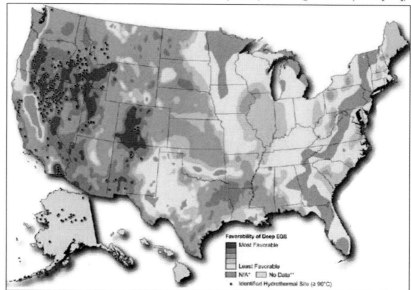

Source: NREL, available at http://www.nrel.gov/gis/images/geothermal_resource2009-final.jpg.

Notes: Map does not include shallow EGS resources located near hydrothermal sites or USGS assessment of undiscovered hydrothermal resources. Source data for deep EGS includes temperature at depth from 3 to 10 km provided by Southern Methodist University Geothermal Laboratory (Blackwell & Richards, 2009) and analysis (for regions with temperatures ≥150«C) performed by NREL (2009). Source data identified hydrothermal sites from USGS Assessment of Moderate- and High-Temperature Geothermal Resources of the United States (2008).

* "N/A" regions have temperatures less than 150«C at 10 km depth and were not assessed for deep EGS potential.

** Temperature at depth data for deep EGS in Alaska and Hawaii not available.

Figure 8. Geothermal Resource of the United States.

EIA estimates that the levelized cost of energy (LCOE) for conventional geothermal electricity ranges from \$92/MWh to \$116/MWh.[50] *Figure 12* provides a comparison of costs for conventional (fossil and nuclear) and renewable electricity generation. NREL has also estimated geothermal electricity LCOE and concluded that, depending on the total amount of capacity installed, geothermal (conventional and EGS) electricity costs could range between \$50/MWh and \$1,200/MWh (2008 US\$).[51]

Based on these cost of energy estimates, NREL indicates that the amount of EGS resource "that can be economically produced is likely much smaller" than the total resource potential.[52]

Hydroelectric

U.S. Resource Estimates

Hydropower is currently the largest source of renewable electricity production in the United States. In 2010, approximately 257,000 GWh was generated from hydropower resources, equal to roughly 7% of total U.S. electricity generation.[53] Hydropower can be generated in many ways. For the purpose of this report, hydropower refers to "conventional" hydropower[54] and does not include hydrokinetic energy, ocean energy, or pumped storage.[55]

Since 1998, several Idaho National Laboratory (INL) reports have estimated the potential to develop new hydropower generation capacity. Three of those reports trace a time-wise increase in the estimates of potential generation capacity: 30 GW (1998),[56] 43 GW (2003),[57] and 60 GW (2006).[58] A review of the INL reports revealed that estimates differed with regard to the hydropower categories included in the calculations.[59] After sorting through the studies and attempting to remove duplicative and non-relevant data, CRS calculated additional hydropower potential to be approximately 65 gigawatts, which equates to approximately 284,700 GWh of additional annual electricity generation potential.[60] A 2007 report by the Electric Power Research Institute (EPRI) estimated that additional hydropower capacity potential was equal to 62.3 gigawatts.[61] However, recently published Oak Ridge National Laboratory (ORNL) resource potential estimates for non-powered dams may increase the total hydropower resource assessment by as much as 12.6 gigawatts.[62]

Figure 9 provides summary information about the location of existing and potential hydroelectricity facilities in the United States.

Technology and Cost Considerations

Hydroelectricity generation in the United States dates back to the 1880s and many technologies are fully commercialized with proven operational performance.[63] However, DOE is pursuing efforts to further improve the performance, economics, and environmental impact of conventional hydropower technologies.[64] Low-head and low-power hydroelectricity technology that might be used in constructed waterways, such as canals, may

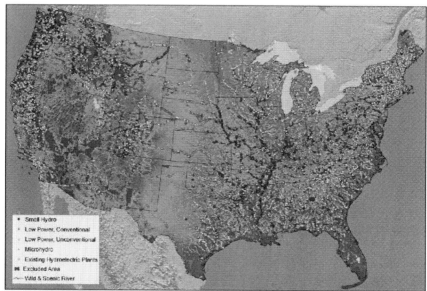

Source: DOE. "Feasibility Assessment of the Water Energy Resources of the United
States for New Low Power and Small Hydro Classes of Hydroelectric Plants,"
DOE, Office of Energy Efficiency and Renewable Energy, Wind and Hydropower
Technologies, January 2006, available at http://www1.eere.energy.gov/
windandhydro/pdfs/doewater-11263.pdf.

Notes: Alaska and Hawaii were included in the DOE study, but were not included in
the accompanying map. DOE study results indicate that Alaska may have the
potential to increase its hydropower capacity by as much as 16 times and Hawaii
was distinguished as the state having the highest concentration (measured as
kilowatt-annual per square mile) of hydropower potential.

Figure 9. Existing and Potential Hydropower Projects in the Lower 48 United States.

require additional research, development, and demonstration before being
commercially available.[65]

Hydroelectricity is generally considered to be one of the lowest-cost
sources of renewable electricity. EIA estimates that the LCOE for new
hydroelectricity plants ranges from $59/MWh and $121/MWh.[66] *Figure 12*
provides a comparison of costs for conventional (fossil and nuclear) and
renewable electricity generation. However, due to their relatively early stage
of development, the cost of electricity from low-head and low-power
technologies remains somewhat uncertain.

Ocean and Hydrokinetic

U.S. Resource Estimates

Ocean-based energy resources come in several forms, including (1) tidal, (2) wave, (3) current, and (4) thermal (also known as Ocean Thermal Energy Conversion or OTEC).[67] Each ocean energy resource is fundamentally different in terms of the amount of available resources, location of the resource, and the conversion technology used to generate electricity. While a limited number of ocean energy resource assessments are available, the Electric Power Research Institute (EPRI) published resource estimates for wave and tidal energy in 2006.[68] DOE has funded several resource assessments that are not yet available.[69] *Table 4* summarizes some of the resource estimates for different categories of ocean energy.

Figure 10 illustrates the location and magnitude of U.S. wave energy resources. The majority of wave energy potential exists off the coasts of Alaska, Hawaii, and west coast states.

Table 4. U.S. Ocean Energy Resource Estimates

	Resource Estimate (GWh/year)		
	Low	High	Resource Assessment Status
Wave	255,000	2,100,000	In 2008, DOE awarded a Marine Energy Grant to EPRI to assess U.S. wave energy resources.
Tidal	n/a	6,600	Georgia Tech Research Corporation was awarded a grant from DOE to assess tidal stream energy production potential.
Current	n/a	n/a	DOE awarded a grant to Georgia Tech Research Corporation to create a database of ocean current energy potential.
OTEC	n/a	n/a	DOE awarded a grant to Lockheed Martin in 2009 to conduct global and domestic ocean thermal resource assessments.

Source: EPRI (Resource Estimates); "Report to Congress on Renewable Energy Resource Assessment Information for the United States," DOE, Office of Energy Efficiency and Renewable Energy, January 28, 2011.

Notes: n/a = not available.

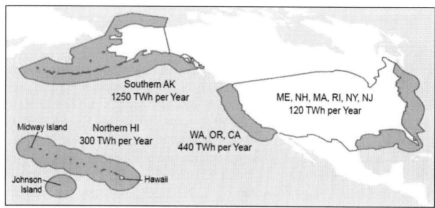

Source: R. Bedard, G. Hagerman, M. Previsic, O. Siddiqui, R. Thresher, and B. Ram,
    "Final Summary Report: Project Definition Study – Offshore Wave Power
    Feasibility Demonstration Project," Electric Power Research Institute, September
    22, 2005, available at http://oceanenergy.epri.com/attachments/wave/reports/ 009_
    Final_Report_RB_Rev_2_092205.pdf.
Notes: TWh = terawatthours. 1 TWh = 1,000 gigawatthours.

Figure 10. U.S. Wave Energy Resources.

Hydrokinetic energy, defined as river in-stream energy for the purpose of
this report, can be extracted from the natural water flow in rivers. The amount
of electricity that can be generated from this energy source is dependent on the
volume and velocity of the water resource. A DOE-funded study by New York
University estimates that approximately 12.5 GW of hydrokinetic power
potential might be possible.[70] Assuming a capacity factor between 30% and
50%, electricity generation potential from hydrokinetic resources may range
from 32,850 GWh to 54,750 GWh.[71]

Technology and Cost Considerations

Ocean and hydrokinetic electricity generation technologies might be
considered "emerging" as they have yet to operate at a significant commercial
scale. Nevertheless, demonstration and commercial deployment of ocean and
hydrokinetic projects is being pursued.[72] Many technology concepts are being
developed and demonstrated, including more than 100 ocean energy devices
worldwide, with approximately 30 under development in the United States.[73]

Given the early developmental status of ocean and hydrokinetic electricity
production technologies, estimating the levelized cost of energy is
challenging.[74]

EIA did not include an LCOE estimate for ocean and hydrokinetic electricity generation as part of the Annual Energy Outlook (AEO) 2011.

Biomass

U.S. Resource Estimates

Accurate estimates for biomass electricity generation potential are somewhat challenging because biomass material (forest, agriculture, solid waste, and landfill gases) can be used in a variety of competing ways to include electricity generation, biofuel production, and space heating for residential and commercial buildings.[75] Also, unlike other renewable energy sources, biomass might be considered a managed resource in that the quantity of biomass material available for electricity generation can go up or down based on changes in management practices.[76] As a result, U.S. biomass electricity generation potential is highly dependent on how much biomass is available and how much biomass material is dedicated for this specific use. In 2009 an estimated 54,493 GWh of electricity was generated from biomass, which represented approximately 1.2% of total U.S. net electricity generation.[77]

According to DOE, approximately 190 million tons of biomass are consumed each year, with roughly 25% to 35% of current biomass consumption being used for electricity generation. DOE analysis and reports indicate that the potential may exist to produce about 1.3 billion tons of biomass annually.[78] However, estimating the amount of electricity that might be generated from biomass depends on the amount of biomass material available and on the portion of that material that might be used for electricity generation. *Table 5* provides an estimate for potential generation if DOE's 1.3 billion ton estimate of biomass production were realized.

Figure 11 indicates the relative concentration of current biomass resources throughout the United States.

Technology and Cost Considerations

Combustion technologies used to convert biomass to electricity are generally considered commercial, and there are approximately 80 operating biomass electricity generation facilities located in the United States.[79] Nevertheless, using biomass as a feedstock for electricity generation can be challenging because each biomass type has different properties, such as water content, ash content, and energy value.[80]

Table 5. Annual U.S. Biomass Electricity Generation Potential

% of 1.3B	10%			50%		100%
tons	Low	High	Low	High	Low	High
Electricity Generation (GWh)	125,730	142,880	628,660	714,390	1,257,330	1,428,780

Source: CRS analysis of scenarios based on, "Biomass as Feedstock for a Bioenergy and Bioproducts Industry:
The Technical Feasibility of a Billion-Ton Annual Supply," U.S. Department of Energy and U.S. Department of Agriculture, April 2005, available at http://www1.eere.energy.gov/biomass/pdfs/final_billionton_vision_report2.pdf.
Notes: Calculations for this table were based on the following: (1) annual tons consumed for electricity generation, (2) energy content (Btu) per ton of biomass, and (3) biomass-to-electricity conversion efficiency. Estimates of annual tonnage were based on the percentages listed in the table (10%, 50%, and 100%). Energycontent per ton of biomass was assumed to be 15 million Btu/ton. Biomass-to-electricity conversion efficiency ranged from 22% to 25%. This range is the reason for "low" and "high" estimates in the table. This conversion efficiency is generally representative of biomass combustion technologies, which have a commercial operating history. Other conversion technologies, such as certain gasification or biological conversion approaches, may have different conversion efficiencies. Since there are competing uses for biomass material, it is unlikely that 100% of the potential biomass resource will be used for electricity generation. The "100%" scenario presented in this table is provided for reference only.

This variability in feedstock quality and characteristics typically must be addressed in order to effectively operate biomass electricity generation equipment.

Further, some biomass materials contain certain alkali metal species, such as sodium and potassium, that can potentially impede the operation of electricity generation equipment.[81]

EIA estimates that the levelized cost of energy for biomass electricity ranges from $99.50 per MWh to $133.40 per MWh. Biomass accumulation and transportation and biomass feedstock quality might be considered key cost drivers that can impact the levelized cost of energy for biomass electricity.[82] *Figure 12* provides a comparison of costs for conventional (fossil and nuclear) and renewable electricity generation.

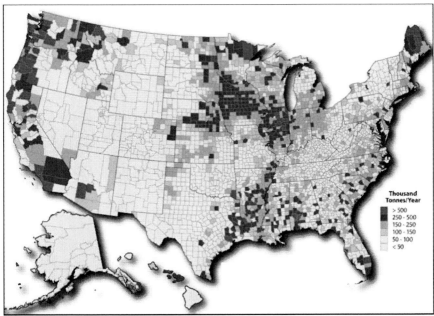

Source: NREL, available at http://www.nrel.gov/gis/images/map_biomass_total_
us.jpg.

Notes: This NREL study estimates the biomass resources currently available in the United States by county. It includes the following feedstock categories: crop residues (five year average: 2003-2007), forest and primary mill residues (2007), secondary mill and urban wood waste (2002), methane emissions from landfills (2008), domestic wastewater treatment (2007), and animal manure (2002). For more information on the data development, please refer to http://www.nrel.gov/docs/fy06osti/39181.pdf. Although the document contains the methodology for the development of an older assessment, the information is applicable to this assessment as well; the difference is only in the data's time period.

Figure 11. U.S. Biomass Resource Availability.

CHALLENGES FOR RENEWABLE ENERGY

Each type of renewable energy technology has certain advantages and disadvantages relative to each other and relative to fossil fuel energy sources. An extensive literature exists on these advantages and disadvantages.[83] This part of the report offers brief observations and discussion of certain challenges that might affect the full development and deployment of renewable electricity generation technologies. Many policies directed at renewable energy deployment are designed to address these challenges.

Cost

Perhaps the most fundamental challenge to the deployment of renewable energy is the cost of generating electricity from renewable sources. Energy producers and consumers seek the lowest-cost energy, and fossil fuels have historically been the lowest-cost sources of energy, either through end-use combustion or through the generation of electricity.

Levelized Cost of Energy (LCOE)

A common metric for measuring the financial cost of electricity production is Levelized Cost of Energy, or LCOE.[84] LCOE calculations are typically expressed in terms of dollars per unit of energy. The most common units of energy used for comparing the LCOE of different energy sources are kilowatthour (kWh) and megawatthour (MWh). LCOE estimates can provide a relative comparison of energy generation costs for different energy sources such as coal, natural gas, wind, solar, and others. However, policy makers may want to exercise caution when reviewing and considering LCOE estimates. Reasons for this caution include the following: (1) no agreed-upon or standardized LCOE calculation methodology exists, and methods can be tailored to skew results in favor of a particular technology or resource, (2) assumptions used to calculate LCOE estimates can have a major impact on calculations results, and (3) LCOE estimates may not reflect the variable time-of-day value of electricity generation. For example, electricity at 2 p.m. may have more value than electricity at 2 a.m. Therefore, it is unlikely that LCOE calculations performed by different organizations will be identical. Understanding the methodology and assumptions used is often critical when considering LCOE estimates.

Although there is no standard LCOE calculation method, two fundamental methods are commonly used. One method uses total life cycle costs (capital, operations and maintenance, etc.) and total life cycle energy production to calculate a $/kWh or $/MWh cost of energy.[85] Another method uses a project cash flow model to calculate equity rates of return based on the price of energy paid to the project. Typically, a target equity rate of return, expressed as a percentage, is established and the price per unit of energy is adjusted in order to reach the equity return target.[86] This cash-flow-based methodology is unique in that it may include specific project finance constraints such as debt service coverage ratios, cash reserves, and other factors that may not be reflected in the cost vs. energy production approach.

Furthermore, differences in several key assumptions can significantly alter calculations of LCOE estimates. Assumptions that can impact LCOE estimates include (1) capital costs, (2) operation and maintenance costs, (3) government incentives, (4) capacity factor, (5) financial structure (debt/equity ratio), (6) financial costs for debt and equity, (7) project lifetime, and (8) technology performance degradation. Several key assumptions must be included in each calculation of LCOE estimates. Since different organizations often use different assumptions, the variation in LCOE estimates is not surprising.[87]

For this report, LCOE estimates from the EIA Annual Energy Outlook (AEO) 2011 were used to compare the cost of new electricity generation for various renewable energy resources. *Figure 12* summarizes EIA's range of LCOE estimates for several technologies.[88]

(2009 $/Megawatthour)

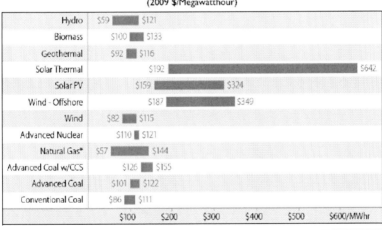

Source: CRS adaptation of EIA's "Levelized Cost of New Generation Resources in the Annual Energy Outlook 2011," available at http://www.eia.gov/oiaf/aeo/ electricity_generation.html.

Notes: EIA LCOE estimates are for new projects that are would be brought on line in 2016. LCOE estimates do not incorporate any federal or state tax incentives.

* The LCOE range for Natural Gas includes four different technologies: (1) conventional combined cycle, (2) advanced combined cycle, (3) conventional combustion turbine, and (4) advanced combustion turbine.

Figure 12. EIA's Levelized Cost of Energy (LCOE) Estimates for New Plants.

It is important to note that EIA LCOE estimates reflect only the projected amount of capacity expected to be added to the electricity generation system during the forecast period. Costs for renewable electricity typically follow a supply curve where costs increase as new capacity is installed, which indicates that the lowest-cost capacity will be added first. Geothermal supply curve estimates provide an example to consider. Figure 13 shows a supply curve for enhanced geothermal electricity costs, developed by NREL. As indicated in the figure, depending on the amount of geothermal capacity installed, the projected LCOE could be as high as $1,000 per MWh. This example of how energy costs can change, as capacity additions increase, further illustrates the importance of understanding all assumptions used for projecting future electricity costs.

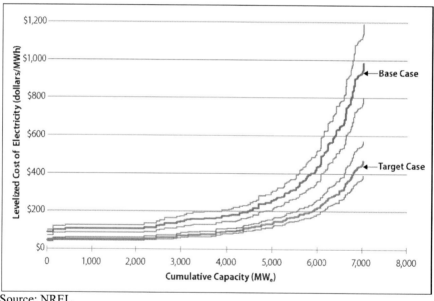

Source: NREL.

Notes: For more information see C. Augustine, K. Young, and A. Anderson, "Updated U.S. Geothermal Supply Curve," National Renewable Energy Laboratory, Conference Paper presented at Stanford Geothermal Workshop, February 1, 2010, available at http://www.nrel.gov/docs/fy07osti/41073.pdf.

Figure 13. NREL Supply Curve for Near-Hydrothermal Field Enhanced Geothermal Systems (EGS) Resource.

Comparing Fossil and Renewable Energy Costs

The current comparatively low cost of fossil fuel energy, as indicated in EIA's LCOE estimates, may not include any costs associated with the external impacts of fossil fuel consumption, which has been stated this way:

> But some energy costs are not included in consumer utility or gas bills, nor are they paid for by the companies that produce or sell the energy. These include human health problems caused by air pollution from the burning of coal and oil; damage to land from coal mining and to miners from black lung disease; environmental degradation caused by global warming, acid rain, and water pollution; and national security costs, such as protecting foreign sources of oil.[89]

Accurately quantifying social costs of fossil energy associated with health problems, climate change, and others can be difficult and complex. As with all cost calculations, assumptions used for estimating social costs can have a dramatic effect on calculation results. Nevertheless, some groups do attempt to place a value on the social costs of fossil energy as an alternative method for comparing the cost of energy from fossil and renewable resources.[90] Furthermore, an Interagency Working Group was created under Executive Order 12866 to estimate the social cost of carbon for regulatory impact analysis.[91] Over time, the costs of mitigating some of these social costs may be placed on the producers or consumers of fossil fuels.

Technology and cost are closely related because renewable energy developers seek technologies that produce energy as inexpensively as possible in order to attain commercially viability. Today, research continues to identify new, more efficient materials and to seek technologies that can be manufactured at lower cost. Improvements continue as new technologies emerge and evolve.

Much of the current R&D on renewable technologies aims to reduce the manufacturing cost and the electricity production cost, thereby making renewable electricity more competitive in the marketplace.

Power System Integration

Connecting renewable electricity generation facilities to the electric power grid can raise potential technical challenges. In particular, a high percentage penetration of variable sources—such as solar and wind—can cause serious power quality and reliability problems. The power system requires constant, 24/7 minute-by-minute monitoring and control. The introduction of variable electricity generation may pose power system reliability challenges associated with moment-bymoment balancing of electricity supply and demand.[92] A recent study by the International Energy Agency (IEA) indicates that many existing power systems currently have infrastructure and processes to manage some degree of variability, and these existing assets could potentially be used to manage variable renewable energy resources.[93] However, not all renewable sources of electricity are classified as variable. Biomass, geothermal, and some hydropower sources have the ability to generate electricity on a consistent and predictable basis. As a result, integrating these renewable sources into the power system may not be difficult. However, the inherently variable nature of wind, solar, and some ocean-hydrokinetic electricity may result in significant power system operational challenges if these variable renewable energy sources achieve a high percentage level of penetration.[94]

DOE funded two studies—the Eastern and Western grid interconnection studies—to evaluate the challenges and opportunities associated with significant penetration of variable renewable sources of electricity.[95] The Eastern Wind Integration and Transmission Study assessed the impacts of integrating wind electricity generation at a 20% to 30% penetration level.[96] The Western Wind and Solar Integration Study evaluated potential power system impacts associated with a 35% penetration comprised of wind (30%) and solar (5%).[97] Both studies concluded that integrating the respective penetration rates of variable renewable electricity is manageable, although accommodating those penetration levels may require large amounts of transmission investment, additional reserve capacity, and modifications to power system operations.

Intermittency and Variability

Some renewable energy sources are intermittent and variable. Geothermal, biomass, and some hydropower energy sources usually can be delivered continuously over time. However, wind power is usable only when the wind blows, solar power is usable only when the sun shines, and some hydroelectric power is usable only when water is available to flow through the turbines, so the production of renewable electricity from those sources varies over a period of minutes, hours, days, or months. In addition, wind speed may vary over a period of seconds, minutes, or hours, and solar energy may vary with cloud cover over a period of minutes or hours. This intermittent and variable nature of renewables contrasts with fossil and nuclear power plants, which produce electricity continuously and uniformly except during times of maintenance, fuel supply disruptions, operational problems, or natural disasters. The intermittency and variability of renewable energy might be partially overcome through the development of advanced storage technologies that provide storage of various quantities of electrical energy for use during renewable energy down time. A wide range of batteries, compressed-air storage, hydrogen generation and fuel cells, and other means of storing and recovering intermittent energy are being studied. Such storage is currently costly, and the combination of renewable electricity generation and reliable storage—or backup reserve capacity from natural gas or other dispatchable sources— will need to be considered by the electrical delivery system in order to maximize the potential contributions of renewable technologies.[98]

Renewable Energy Footprint and Land-Use

Although the amount of renewable energy available from the sun, wind, and water may seem unlimited, the land available for energy development is potentially limited by a number of factors. As mentioned above, these sources are dispersed, and technologies are required to convert the natural form of energy into electricity. For this reason, certain renewable energy technologies available today require large areas of land—a large footprint—for each unit of energy produced.

Figure 14 displays different estimates of the land-use intensities of several energy production technologies. Estimates of the land-use intensity for renewable and nonrenewable sources of energy vary significantly, depending on a number of assumptions. To date, there is no standard methodology to produce these estimates. For example, the extent to which an energy site exclusively "uses" an amount of land is debatable. Only a small portion of area within a wind energy site is actually occupied by the turbines, so remaining land could potentially be—and often is—dedicated to other uses. In contrast, fields of energy crops to be burned in the production of electricity will fully occupy their allotted area. Further, energy production for renewables varies substantially with geography. A solar photovoltaic plant of a certain capacity will require less land if located in a region with more intense sunlight. In addition, it is difficult to compare land-use intensities for renewable energy technologies with those of fossil fuel technologies. For example, for fossil fuels, calculations of land-use intensity may include the power plant footprint, plus mining or production area, plus areas occupied by transportation and logistics infrastructure. Thus, the footprint for natural gas may include the gas power plant, but also the areas occupied by gas wells, the roads that connect the gas wells, and the pipelines that transport the gas to market. Also, the areal extent of infrastructure may not fully represent the impact on the landscape. The degree to which such infrastructure divides or dissects ecosystems may also be an important consideration. [99]

The electric energy production technologies with the greatest land-use intensity (amount of land per unit of electrical energy produced) are biomass, wind, hydropower, and solar photovoltaic. Land-use intensities of natural gas, coal, geothermal, and nuclear power are likely significantly smaller than those of other forms of energy production. As demand grows for utility-scale installations of renewable energy, pressure will grow to integrate energy policy with land-use policy.[100] The integration of distributed generation technologies, such as rooftop solar, into existing building structures will help mitigate land-use issues, but there will likely remain a strong need for utility-scale renewable energy installations.

Transmission Availability and Access

Though renewable energy technologies may be used across most of the nation, optimized use of renewable energy must accommodate certain geographic controls. The wind energy resource is richest in coastal areas and the Midwest. Solar energy is optimal in the relatively cloudless southwestern United States. Hydroelectric power has historically been best deployed on large rivers with steep gradients. These geographic concentrations of renewable energy sources often mean that the energy may be optimally produced far from the existing energy demand centers, which are the large cities of the east and west coasts, upper Midwest, and South. Thus, large-scale deployment of renewable energy technologies will likely be accompanied by the need for new electricity transmission infrastructure from the new regions of energy supply to the demand centers.[101] For example, the NREL Eastern Interconnection Report concluded that 20% to 30% wind generation is feasible, but would require "significant expansion of the transmission infrastructure."[102]

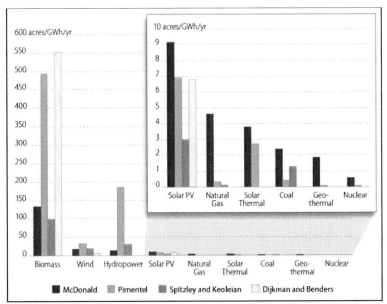

Source: CRS analysis of the following reports:
- McDonald RI, Fargione J, Kiesecker J, Miller WM, Powell J(2009) Energy Sprawl or Energy Efficiency: Climate Policy Impacts on Natural Habitat for the United States of America. PLoS ONE 4(8): e6802. doi:10.1371/journal.pone.0006802, Figure 3.
- David Pimentel et al., "Renewable Energy: Current and Potential Issues," BioScience, vol. 52, no. 12 (December 2002), pp. 1111-1120.
- David V. Spitzley and Gregory A. Keoleian, Life Cycle Environmental and Economic Assessment of Willow Biomass Electricity: A Comparison with Other Renewable and Non-Renewable Sources, Center for Sustainable Systems, Report No. CSS04-05R, Ann Arbor, MI, March 25, 2004 (revised February 10, 2005).
- T.J. Dijkman and R.M.J. Benders, "Comparison of renewable fuels based on their land use using energy densities," Renewable and Sustainable Energy Reviews, vol. 14 (2010), pp. 3148-3155.
Notes: GWh = gigawatthours, yr = year; Acres per gigawatthour per year (GWh/yr) is the metric used to compare results from the respective reports. GWh/yr indicates the amount of land required to generate a certain amount of electricity, in this case a gigawatthour. Some studies report land use per unit of capacity, which might be reported as acres per gigawatt (GW). Land use per capacity is somewhat misleading because each energy technology has a different capacity factor, meaning that operational hours for each technology will vary over the course of a year. Reflecting land use as a function of electricity generation takes into account capacity factor differences.

Figure 14. Land-Use Intensity for Various Forms of Energy Production.

Not only must a new installation of renewable energy technology be connected to the grid, but the new transmission infrastructure must be sized to the maximum rate of electricity flow even though it may flow intermittently at that rate.

Materials and Resources

While renewable energy sources may provide a virtually infinite supply of energy, building and installing the equipment necessary to convert renewable energy into usable electricity may require significant quantities of materials and other natural resources. For example, wind turbine manufacturing requires a number of materials and resources, the most critical being steel, fiberglass, resins, blade core materials, permanent magnets, and copper.[103] Current solar photovoltaic technologies require materials such as silicon, cadmium, tellurium, silver, and others.[104] Large-scale wind and solar deployment would raise demand for these materials, which in turn may impact their respective prices. This potential price impact may be an important consideration, since the cost of renewable electricity generation is highly correlated with the cost of the energy conversion system (i.e., wind turbines, solar panels, etc.).

Environmental Impact and Aesthetic Concerns

Capturing and converting any energy source—including renewable energy—will have some degree of impact on the environment. Land use and habitat disturbance are potential environmental issues for wind and solar electricity projects. Installation of wind turbines has already attracted attention because of bird mortality, noise, and resulting NIMBY[105] attitudes. Some CSP technologies may require vast amounts of water, although dry-cooling CSP technologies, with lower efficiencies, are available.[106] Water use, land subsidence, and seismicity may need to be addressed by geothermal power plants. Hydropower and ocean-hydrokinetic electricity generation systems may result in water quality degradation, ecosystem disruption, and animal mortality. Biomass projects impact the environment through emissions such as nitrogen oxides (NO_x), carbon dioxide (CO_2), and others, as well as land use changes associated with producing biomass feedstock.[107] These, and other, potential environmental impacts may need to be considered as policy makers look to balance the desire to increase electricity production from renewable sources of energy with environmental objectives.

Infrastructure Requirements

All forms of energy production and delivery require some form of infrastructure. Coal is delivered by an extensive network of railroads, and natural gas is delivered via a large network of pipelines. In addition to new transmission requirements, some renewable energy sources may require investments in specialized infrastructure in order to provide a source of renewable electricity. One example is offshore wind energy. Specialized vessels, purpose-built portside infrastructure, undersea electricity transmission lines, and grid interconnections will likely be required to support offshore wind development. According to DOE, "these vessels and this infrastructure do not currently exist in the U.S."[108] Such specialized infrastructure requirements may also be a consideration for policy decisions associated with certain other types of renewable energy.

Technology Development and Commercialization

Some renewable electricity generation technologies are not yet commercially available. Private and public investments are being made in renewable electricity generation technologies, to include venture capital firms, private and public corporations, and the U.S. DOE through its Advanced Research Projects Agency—Energy (ARPA-E) program office. While ARPA-E and DOE's Office of Energy Efficiency and Renewable Energy provide funds to support technology R&D, concept demonstrations, and technology performance optimization, bridging the gap between these activities and commercialization may require significant amounts of funding. Commonly known as the commercialization "valley of death," several additional market development activities that might include technology performance characterization and validation, operational reliability assessments, accurate quantification of maintenance and operations costs, etc., may be necessary in order for new technologies to qualify for private equity and bank/debt finance in support of commercial projects. Obtaining the funds necessary to commercialize new technologies can be difficult and costly.[109]

Policy and Regulatory Challenges

Certain federal and state-level policies have served to stimulate growth of renewable electricity generation. Federal policies such as production and investment tax credits for certain renewable energy property, along with other tax-favored finance options, have created financial incentives for building and operating renewable electricity generation projects.[110] Other federal financial incentives are also available for renewable energy.[111] Furthermore, the American Recovery and Reinvestment Act (ARRA) provided new policies such as the Section 1603 cash grant option for renewable electricity generation projects and the Section 1705 Loan Guarantee Program that provides government-backed debt financing for certain renewable energy projects.[112] State-level policies, such as renewable portfolio standards, have served to create market demand for renewable electricity.[113] Furthermore, there is some interest in establishing a federal renewable or clean energy standard, which may create additional demand for renewable electricity generation.[114]

Federal policies that support renewable electricity generation typically are available for a defined period of time, at the end of which the policies expire. Some in the renewable electricity industry argue that the sudden expiration of certain federal policies has resulted in market uncertainty and downward pressure on renewable electricity market growth.[115] The historical start-stop nature of federal policies may be challenging to the renewable energy industry due to a lack of long-term financial certainty for renewable electricity generation projects. On the other hand, some policy makers may not wish to create a policy environment that results in a renewable energy industry that is dependent on federal financial incentives. Balancing policy objectives that might stimulate a solid base for renewable electricity, while at the same time eliminating a dependency on federal subsidies, may be a consideration for policy makers.

RELATED ISSUES

Energy Efficiency and Curtailment

Although this report does not provide a detailed analysis of energy efficiency and conservation, it is widely acknowledged that both energy efficiency (doing as much or more with less energy and eliminating waste) and curtailment of demand (doing less with less energy) provide enormous opportunities for reducing or controlling the energy resources of the nation. By addressing the demand side of the energy equation, as well as the supply side, the United States can extend the energy resources that it consumes. Efficiency and demand curtailment will not, by themselves, meet the demand for energy in the future, but these strategies will likely reduce the amount of new energy needed.[116]

The benefits of more efficient use of energy are being sought by a wide range of citizens, homeowners, manufacturers, and governments. Lower costs, reduced greenhouse gas emissions, and reduced need for expansion of supply are key motivators to increase energy efficiency and conservation. Energy efficiency can be measured for individual devices such as appliances, automobiles, and light bulbs, but derivative indicators are used to measure levels and trends in energy efficiency at a national level. The most common national indicator of energy efficiency and curtailment is energy intensity.[117] Energy intensity is measured in units of energy per dollar of Gross Domestic Product (GDP). As *Figure 14* shows, the energy intensity of the United States has been dropping steadily for decades, despite the steady growth in total energy consumption. There are many reasons for this trend, of course, including a gradual change from a manufacturing economy to a more service-oriented economy, but ongoing efforts to promote energy efficiency and conservation are clearly succeeding in the United States. For example, one study estimates that improving the energy efficiency of buildings in the United States could save $170 billion per year in energy costs through 2030.[118] Numerous other opportunities exist for improving efficiency or curtailment in energy use in the United States.[119]

Biofuels

Biofuels are liquid fuels produced from plant materials, which makes them a renewable commodity. The major biofuels are fuel ethanol and biodiesel, though other kinds of alcohols and hydrocarbons can also be synthesized from biological materials. Both fuel ethanol and biodiesel are currently used primarily as blending agents with conventional gasoline and diesel fuel, though both can conceivably be used in their pure form with some modifications to engine fuel systems.[120] Unlike the other kinds of biomass discussed above, liquid biofuels are normally used as transportation fuels and are not used to generate electricity. Liquid biofuels are important because certain forms of transportation such as aircraft and heavy trucks cannot easily

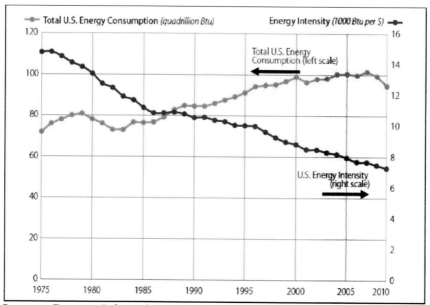

Source: Energy   Information   Administration,   http://www.eia.gov/emeu/aer/pdf/
           pages/sec1_13.pdf.
Note: Energy intensity is the total primary energy consumption per real dollar of Gross
      Domestic Product.

Figure 15. Total U.S. Energy Consumption and Energy Intensity, 1975-2009.

be converted to electricity or other propulsion technologies. In 2009, the
United States consumed 99 million gallons of fuel ethanol as an 85% blend
(E85), 10.7 billion gallons of fuel ethanol as a 15% blend (E15) in gasoline,
and 316 million gallons of biodiesel.[121]

ADDITIONAL CONSIDERATIONS FOR RENEWABLE
ELECTRICITY IN THE UNITED STATES

The Scale of U.S. Energy Consumption

One important aspect of the expansion of renewable forms of energy,
often overlooked or under-appreciated, is the scale or magnitude of energy use
in the United States. It is not only *what kind* of energy is used, but *how much*
energy the United States uses on a daily, monthly, and annual basis. By any
measure, the amounts of energy used by the United States are prodigious, and

replacing a significant proportion of fossil fuels with renewable forms of energy would be a formidable task. Alternatives to fossil fuels must be produced on a very large scale and must be available to all parts of the nation to provide the enormous and increasing amounts of energy demanded by the U.S. economy. Whether individual renewable energy installations are large (utility-scale) or small (distributed), the total combined output must accommodate the very large—and increasing—demand for energy. Current electricity generation is dominated by coal, natural gas, and nuclear (see Table 6). Only 8% of total energy use in the United States is renewable, and 53% of that is for electricity generation. In 2009, total U.S. energy use was 94.6 quadrillion Btu, and renewable electricity accounted for about 4 quadrillion Btu.[122] Therefore, any serious proposal to displace fossil fuels with renewable energy must include massive growth in renewable energy technology deployment.

Table 6. Total U.S. Electricity Generation, By Source, 2009

Generation fuel	GWh	%
Coal	1,755,904	44.45
Petroleum	38,937	0.99
Natural Gas	920,979	23.31
Other Gases	10,632	0.27
Nuclear	798,855	20.22
Hydroelectric Conventional	273,445	6.92
Wind	73,886	1.87
Solar Thermal and Photovoltaic	891	0.02
Wood and Wood Derived Fuels	36,050	0.91
Geothermal	15,009	0.38
Other Biomass	18,443	0.47
Pumped Storage	-4,627	-0.12
Other	11,928	0.30
All Energy Sources	3,950,332	100.00

Source: EIA, http://www.eia.doe.gov/cneaf/electricity/epa/epaxlfilees1.pdf.
Notes: Electricity generation from pumped storage is negative since pumping water into a storage reservoir requires more electricity than that generated when the stored water is used to operate a turbine. Pumped storage projects are typically based on opportunities to pump water into a reservoir when electricity prices are low (typically at night), then use the stored water to generate electricity when prices are high (typically during peak demand hours).

Relationship between Renewable Electricity and Imported Energy

Petroleum consumption may be displaced by the production of biofuels, but most renewable energy technologies are designed to generate electricity. Therefore, the use of renewable energy to generate electricity in today's U.S. market would displace only those fossil fuels that are used to generate electricity, and the United States uses almost no imported fossil fuels to generate electricity. For example, the U.S. transportation system is 94% reliant on petroleum (*Figure 1*), and the use of renewable electricity for transportation might require increased electrification of the transportation system. Consequently, the only way that increasing production of renewable electricity would affect oil imports is if the U.S. transportation system is electrified so that domestically generated electricity substitutes for oil. Likewise, any process in which the burning of natural gas is used for direct heating would need to be electrified in order for renewable energy to substitute. More than 93% of U.S. coal consumption is used to generate electricity, so adopting renewable energy sources to generate electricity could potentially reduce demand for coal, but would have no effect on energy imports because virtually all of U.S. coal is produced domestically.

International Renewable Electricity Markets

Recent news reports emphasize how successful China and other nations have become in developing and deploying renewable energy technologies. Indeed, China is constructing impressive amounts of renewable energy installations, but the United States remains one of the world leaders in renewable energy capacity and deployment. For example, *Table 7* shows that the United States leads the world in installed non-hydropower renewable electricity generation capacity, biomass power, and geothermal power. While the United States ranked second, behind China, in total wind power capacity, the United States ranked first in 2010 in terms of *operational* wind power capacity.[123] In 2009, the United States generated more electricity from non-hydro renewable energy sources than any other country in the world. However, when hydropower is included, China led the world in terms of total renewable electricity generation (see Figure 16).

**Table 7. Existing Renewable Energy Capacities at the End of 2010
(Country ranking for selected categories)**

Rank	Renewables power capacity (not including hydro)	Renewables power capacity (including hydro)	Wind power	Biomass power	Geothermal power	Solar PV	Solar hot water/heat
1	United States	China	China	United States	United States	Germany	China
2	China	United States	United States	Brazil	Philippines	Spain	Turkey
3	Germany	Canada	Germany	Germany	Indonesia	Japan	Germany
4	Spain	Brazil	Spain	China	Mexico	Italy	Japan
5	India	Germany/ India	India	Sweden	Italy	United States	Greece

Source: REN21. 2011. Renewables 2011 Global Status Report (Paris: REN21 Secretariat), available at http://www.ren21.net/REN21Activities/Publications /GlobalStatusReport/GSR2011/tabid/56142/Default.aspx.

(Selected Countries)

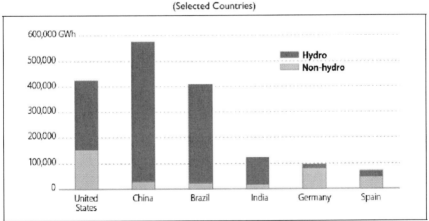

Source: Energy Information Administration, International Energy Statistics, http://www.eia.gov/cfapps/ ipdbproject/IEDIndex3.cfm?tid=6&pid=29&aid=12.

Notes: Non-hydro includes generation from wind, solar, geothermal, tide and wave, and biomass and waste.

Figure 16. Total Net Renewable Electricity Generation, 2009.

FUTURE TRENDS IN RENEWABLE ELECTRICITY

The potential for renewable electricity generation in the United States is very large, yet current use of renewable energy for electricity production is relatively modest, constituting only 11% of total electricity generation and 8% of total energy consumption. Based on the current status of renewables in the United States, policy makers may consider some key questions about the future of renewable energy:

- Should the United States actively seek greater use of renewable energy to supply electricity, or should the energy and electricity markets be allowed to work without further interference with the existing structure of subsidies and incentives?
- If greater use of renewable energy for electricity is desired, what are the key barriers or actions that should be addressed by federal policy?

Future trends in renewable electricity will depend heavily on the cost of both renewable technologies and fossil fuel costs, and on government incentives for renewable energy. In the absence of subsidies for renewable electricity technologies, and in the absence of accounting for external costs of using fossil fuel combustion to generate electricity, several renewable electricity technologies are currently not commercially viable, or only marginally so. Reference case projections by EIA of growth in wind and solar electricity to 2035 are predicated on the use of renewable portfolio standards, renewable fuel standards, and subsidies in the tax code.[124] With low coal and natural gas prices, and high renewable energy technology costs, and the absence of regulation or subsidies, renewable electricity may not increase significantly. Without some form of carbon pricing or other consideration of the externalities of fossil fuel combustion, the United States may remain in an era of relatively low-cost fossil fuel electricity for decades.

However, policy makers may decide that growth in renewable electricity is desirable because of concerns about greenhouse gas emissions and climate change, because fossil fuel supplies are ultimately finite, and because of a desire to position the United States as a global leader for renewable energy technology and manufacturing. Renewables could be made more cost competitive by means of improved renewable technologies or revised cost of carbon-based fuels, but financial or regulatory incentives may be required to make certain renewable sources more economically viable in the short term.

In the event that levelized costs of renewable electricity become competitive with those of fossil fuel electricity, the additional issues of intermittency/variability, land-use and footprint, the need for additional transmission, plus other resource and environmental impacts of renewable electricity will need to be addressed by local, state, and federal officials and policy makers.

CONCLUSION

Cumulative U.S. renewable electricity generation capacity more than doubled from 2006 to 2010, increasing from approximately 22 GW to nearly 55 GW.[125] In 2010, renewable sources of energy provided approximately 11% (7% from hydropower and 4% from other renewables) of total net electricity generation and the EIA AEO 2011 reference case projects that renewable electricity generation will increase to between 14% and 15% by 2035.126 The renewable electricity generation research conducted for this report indicates that the potential may exist for renewable energy sources to make a sizeable contribution toward total U.S. electricity generation demand. However, renewable electricity generation will likely encounter serious challenges, issues, and barriers as technologies and projects look to realize large-scale deployment. As Congress evaluates various energy policy objectives, policy makers may move to holistically evaluate the potential intended benefits, such as emissions reduction and job creation, with potential risks and consequences, such as electricity cost/price increases and electricity delivery reliability issues associated with increasing renewable electricity generation.

End Notes

[1] Energy Information Administration, Annual Energy Review 2009, http://www.eia.doe.gov/ totalenergy/data/annual/ pdf/pecss_diagram_2009.pdf.

[2] Decreasing U.S. reliance on foreign oil is not included here because the focus of this report is on electricity generation. Petroleum contributed 1% of electricity generation in 2009. Based on current U.S. energy infrastructure, adding additional renewable electricity generation capacity will have a negligible, if any, impact on U.S. oil import dependency. However, electrification of the transportation fleet could potentially result in decreasing total U.S. oil demand. Renewable electricity generation combined with electric vehicle market penetration could potentially result in lower oil import requirements.

[3] Biomass and biofuels release CO_2 during combustion, but are considered by some to have zero net emissions because the CO_2 released was taken up from the atmosphere to grow the

plants. However, there is debate about biomass being considered carbon neutral. For more information see CRS Report R41603, *Is Biopower Carbon Neutral?*, by Kelsi Bracmort.

[4] CRS is aware that the National Renewable Energy Laboratory (NREL) is in the process of publishing an analysis about U.S. renewable electricity generation potential. However, the NREL work was not available to influence the research for this report.

[5] Energy Information Administration, http://www.eia.gov/energyexplained/index.cfm? page= renewable_home.

[6] Energy Information Administration, http://www.eia.gov/state/state-energy-profiles-analysis. cfm?sid=WA.

[7] A British thermal unit (Btu) is a measure of the energy (heat) content of fuels. It is the quantity of energy (heat) required to raise the temperature of 1 pound of liquid water by 1°F at the temperature that water has its greatest density (approximately 39°F), http://www.eia. doe.gov/energyexplained/index.cfm?page=about_btu.

[8] A kilowatthour is equal to one thousand watthours. 1 kWh = 3,412 Btu.

[9] A gigawatthour is equal to one billion watthours.

[10] For more information on U.S. fossil fuel resources, see CRS Report R40872, *U.S. Fossil Fuel Resources: Terminology, Reporting, and Summary*, by Gene Whitney, Carl E. Behrens, and Carol Glover.

[11] "Assessment of Moderate- and High-Temperature Geothermal Resources of the United States," U.S. Geological Survey, 2008, available at http://pubs.usgs.gov/fs/2008 /3082/pdf/ fs2008-3082.pdf, and "The Future of Geothermal Energy: Impact of Enhanced Geothermal Systems (EGS) on the United States in the 21st Century," Massachusetts Institute of Technology, 2006, available at http://geothermal.inel.gov/publications/future_of_ geother mal_energy.pdf.

[12] Resource estimate changes are not unique to renewable energy. Fossil fuel estimates typically change in response to technology, economic conditions, improved data sets, etc. Shale gas in the United States is an example of how resource estimates can change over time.

[13] For more information about the Annual Energy Outlook reference case see, Energy Information Administration, "Annual Energy Outlook 2011," Report Number: DOE/EIA-0383(2011), April 2011, available at http://www.eia.gov/forecasts/aeo/pdf/0383(2011).pdf.

[14] In February 2010, NREL and AWS Truepower released estimates for windy land area and wind energy potential for the 48 contiguous United States. A revision to these estimates that includes data for Alaska and Hawaii was released in April 2011. This is the first comprehensive update of wind energy potential since 1993. The NREL AWS study evaluates three gross (no system losses included) capacity factor assumptions (30%, 35%, and 40%) and two hub heights (80 meters and 100 meters). NREL/AWS also considered certain land area exclusions such as parks, urban areas, and others. Study results, maps, and data tables available at http://www.windpoweringamerica.gov/wind_maps.asp

[15] NREL's offshore wind study does not take into account any potential area exclusions. The offshore wind study also does not include estimates for Florida, Mississippi, Alabama, or Alaska. For more information see Marc Schwartz, Donna Heimiller, Steve Haymes, and Walt Musial, "Assessment of Offshore Wind Energy Resources for the United States," National Renewable Energy Laboratory, June 2010, available at http://www.windpower ingamerica.gov/pdfs/offshore/offshore_wind_resource_assessment.pdf.

[16] Gigawatthour estimates for offshore wind energy resources were calculated by CRS by applying average capacity factor assumptions of 30% and 40% to megawatt installed capacity estimates from NREL.

[17] "A National Offshore Wind Strategy: Creating an Offshore Wind Energy Industry in the United States," U.S. Department of Energy, Energy Efficiency and Renewable Energy, February 2011, available at http://www1.eere.energy.gov/windandhydro /pdfs /national_ offshore_wind_strategy.pdf.

[18] For the purpose of this report, "commercially available" refers to renewable electricity generation technologies that have achieved an adequate amount of operational time that allows for performance validation, accurate reliability assessments, and an understanding of actual operations and maintenance requirements. These commercialization parameters are typically validated by an independent engineering firm. This independent validation is typically necessary for technologies, and projects that use these technologies, to obtain debt and equity for project development. Furthermore, commercially available technologies typically have an established supply chain of companies that can provide equipment to meet certain project and technology performance specifications.

[19] American Wind Energy Association (AWEA) U.S. Wind Industry Annual Market Report Year Ending 2010.

[20] More information on DOE's Wind Power program is available at http://www1.eere. energy. gov/windandhydro/wind_power.html.

[21] Levelized Cost of New Generation Resources in the Annual Energy Outlook 2011, U.S. Energy Information Administration, available at http://www.eia.gov/oiaf/ aeo/electricity _generation.html.

[22] An overview of solar resources is provided by the Department of Energy at http://www.eere. energy.gov/basics/renewable_energy/solar_resources.html.

[23] For a comprehensive summary of current solar resource assessment information, see "Report to Congress on Renewable Energy Resource Assessment Information for the United States," U.S. Department of Energy, Office of Energy Efficiency and Renewable Energy, 2011.

[24] The approach used to quantify annual solar electricity generation potential was based on input from experts at NREL and Sandia National Laboratory. If different methodologies for calculating solar resource potential, such as quantifying the total amount of solar radiation exposure on the surface area of the United States, are employed, results may be different (likely much higher) than the summary data presented in this section.

[25] For an overview of CSP technology see http ://www.eere. energy.gov/basics/renewable_ energy/csp.html. For an overview of PV technology see http://www.eere.energy. gov/basics /renewable_energy/photovoltaics.html.

[26] Three solar radiation components are typically measured and reported: (1) direct beam solar radiation, (2) diffuse solar radiation, and (3) global solar radiation (the sum of direct beam and diffuse). CSP systems are able to use only direct beam solar radiation. PV systems are able to use both direct beam and diffuse radiation. More detail regarding solar radiation components is available at http://www.eere.energy.gov/basics/renewable_energy/solar_ esources.html.

[27] Solar energy system configurations may include (1) south-facing flat-plate collectors at various tilt angles, (2) oneaxis flat-plate tracking, (3) two-axis flat-plate tracking collectors, and (4) direct-beam one and two-axis tracking concentrating collectors. For more information regarding solar system configurations see, D. Renne, R. George, S. Wilcox, T. Stoffel, D. Myers, and D. Heimiller, "Solar Resource Assessment," National Renewable Energy Laboratory, February 2008, available at http://www.nrel.gov/docs/fy08osti/ 42301.pdf.

[28] For a description of various CSP technologies, see http:// www. solarpaces. org/ CSP _ Technology/csp_technology.htm. For a description of various PV technologies, see http://solarbuzz.com/going-solar/understanding/technologies.

[29] Tom Mancini, "CSP Overview," Sandia National Laboratories.

[30] *Ibid.* Filters applied to derive these analysis results include (1) sites with >6.75 kwh/m2/day direct normal insolation, (2) excluding environmentally sensitive lands, major urban areas, etc., (3) removing land with slope >1%, 4) only including contiguous areas >10km2. Changing these filters (i.e. reducing the direct normal insolation threshold) would yield different results.

[31] *Ibid.* Arizona, California, Colorado, Nevada, New Mexico, Texas, and Utah.

[32] P. Denholm and R. Margolis, "Land-use requirements and the per-capita solar footprint for photovoltaic generation in the United States," Energy Policy 36, 3531-3543, 2008.

[33] For more information see P. Denholm and R. Margolis, "Impacts of Array Configuration on Land-Use Requirements for Large-Scale Photovoltaic Deployment in the United States," NREL, Conference paper presented at SOLAR 2008—American Solar Energy Society (ASES), May 3-8, 2008.

[34] Telephone interview with Robert Margolis at NREL.

[35] P. Denholm and R. Margolis, "Supply Curves for Rooftop Solar PV-Generated Electricity for the United States," National Renewable Energy Laboratory, November 2008, available at http://www.nrel.gov/docs/fy09osti/44073.pdf.

[36] The remaining 2% of installed capacity consists of power tower, linear fresnel, and dish-stirling technologies. See Tom Mancini, "CSP Overview," Sandia National Laboratories.

[37] For more information about crystalline solar cells, see Y.S. Tsuo, T.H. Wang, and T.F. Ciszek, "Crystalline-Silicon Solar Cells for the 21st Century," NREL, May 1999, available at http://www.nrel.gov/docs/fy99osti/26513.pdf.

[38] More information about DOE CSP R&D programs and projects is available at http:// www1.eere.energy.gov/solar/csp_program.html. More information about DOE PV R&D programs and projects is available at http://www1.eere.energy.gov/solar/ photovoltaics_program.html.

[39] For more information on electrical energy storage, see "Energy Storage: Program Planning Document," Department of Energy, Office of Electricity Delivery and Energy Reliability, February 2011, available at http://www.oe.energy.gov/DocumentsandMedia/OE_Energy_Storage_Program_Plan_Feburary_2011v3.pdf. For more information on "smart-grid," see "The Smart Grid: An Introduction," Department of Energy, available at http://www.oe.energy.gov/DocumentsandMedia/ DOE_SG_ Book_Single_Pages(1).pdf.

[40] For more information on EIA assumptions and calculation methodology see http://www.eia.gov/oiaf/ aeo/electricity_generation.html.

[41] More information about DOE's SunShot initiative can be found at http://www1.eere.energy.gov/solar/sunshot/.

[45] Phone interview with Chad Augustine at NREL.

[46] Geothermal Energy Association, more information available at http://geo-energy.org /Basics.aspx.

[47] Geothermal Energy Association, more information available at http://geo-energy.org /plants.aspx.

[48] An overview of Enhanced Geothermal Systems (EGS) technology is available at http://www1.eere.energy.gov/geothermal/pdfs/egs_basics.pdf.

[49] The Department of Energy has established EGS commercialization programs. More information available at http://www1.eere.energy.gov/ geothermal/ enhanced_ geothermal_ systems.html.

[50] For more information on EIA assumptions and calculation methodology see http://www. eia.gov/oiaf/aeo/electricity_generation.html.

[51] C. Augustine, K. Young, and A. Anderson, "Updated U.S. Geothermal Supply Curve," National Renewable Energy Laboratory, Conference Paper presented at Stanford Geothermal Workshop, February 1, 2010, available at http://www.nrel.gov/ docs/fy07osti/ 41073.pdf. NREL LCOE estimates are in 2008 US$.

[52] Ibid.

[53] Energy Information Administration, more information available at http://www.eia.gov/ cneaf/ solar.renewables/page/hydroelec/hydroelec.html.

[54] "Conventional" hydropower resource assessments typically include large hydropower dams, increasing capacity at existing facilities, non-powered dams, small hydro, and low-power hydro. Pumped storage hydroelectricity generation potential is not included in the resource estimates included in this report.

[55] Pumped storage generation potential was not included in the resource assessment literature reviewed for this report. However, pumped-storage projects are being developed and the Federal Energy Regulatory Commission (FERC) has issued pre-permits for about 33 gigawatts of pumped storage capacity (see http://www.ferc.gov/industries/hydropower/gen-info/licensing.asp). The business case for pumped storage might be viewed as an arbitrage opportunity whereby water is pumped to a reservoir when energy prices are low, and the stored water is used to generate electricity when energy prices are high or an opportunity exists to receive a financial premium for stand-by or firm power. According to EIA, more energy is required to pump water into a storage reservoir than is generated when electricity is produced by releasing the stored water. However, pumped storage facilities can provide valuable ancillary on-demand energy production services for electricity grid operators. For more information see http://www.ferc.gov/industries/hydropower/ gen-info/regulation/ pump.asp.

[56] A. Conner, J. Francfort, and B. Rinehart, "U.S. Hydropower Resource Assessment Final Report," Idaho National Engineering and Environmental Laboratory, December 1998, available at http://hydropower.inl.gov/resourceassessment/pdfs/doeid-10430.pdf.

[57] D. Hall, R. Hunt, K. Reeves, and G. Carroll, "Estimation of Economic Parameters of U.S. Hydropower Resources," Idaho National Engineering and Environmental Laboratory, June 2003, available at http://hydropower.inl.gov/resourceassessment/pdfs/project_report-final_with_disclaimer-3jul03.pdf.

[58] D. Hall, K. Reeves, J. Brizzee, R. Lee, G. Carroll, and G. Sommers, "Feasibility Assessment of the Water Energy Resources of the United States for New Low Power and Small Hydro Classes of Hydroelectric Plants," Idaho National Laboratory, January 2006, available at http://hydropower.inl.gov/resourceassessment/pdfs/main_report_appendix_a_final.pdf. Note: This report quantified hydropower resource potential as megawatts-annual (MWa) based on a 50% capacity factor assumption. As a result, CRS had to convert MWa estimates to megawatts (MW) in order to have resource estimates on an equivalent basis.

[59] The primary difference between the reports was the inclusion of low-power (<1MW) hydropower resources in the INL 2006 report.

[60] Electricity generation potential assumes a 50% capacity factor.

[61] "Assessment of Waterpower Potential and Development Needs," Electric Power Research Institute, 2007, available at http://www.aaas.org/spp/cstc/docs/07_06_1ERPI_report.pdf.

[62] Presentation by Brennan T. Smith to the National Hydropower Association Annual Conference, "U.S. Hydropower Fleet and Resource Assessments," Oak Ridge National Laboratory, April 5, 2011, available at http://hydro.org/wpcontent/ uploads/2011/04/ Brennan-Smith-PPT_NHA_April2011_Final.pdf.

[63] For more information on the history of hydroelectricity in the United States see http://www1. eere.energy.gov/windandhydro/hydro_history.html.For an overview of hydroelectricity technologies see CRS Report R41089, *Small Hydro and Low-Head Hydro Power Technologies and Prospects*, by Richard J. Campbell.

[64] For more information, see http://www1.eere.energy.gov/windandhydro/printable_ versions/ hydro_advtech.html.

[65] DOE, in April 2011, announced $10.5 million of funding for small hydropower technologies that could be deployed in constructed waterways. For more information see http://www.energy.gov/news/10255.htm.

[66] See EIA "Levelized Cost of New Generation Resources in the Annual Energy Outlook 2011," available at http://www.eia.gov/oiaf/aeo/electricity_generation.html.

[67] For more information regarding these energy production approaches, see "Ocean Energy Technology Overview," Department of Energy, Office of Energy Efficiency and Renewable Energy, July 2009, available at http://www1.eere.energy.gov/femp/pdfs/44200.pdf. Osmotic, or salinity gradient, power is another possible source of ocean energy. However, this energy production source has not yet been explored or analyzed in great detail. Background on osmotic power is available at http://en.wikipedia.org/wiki/Osmotic_power.

[68] For more information about EPRI's wave energy resource assessment, see http:// oceanenergy. epri.com/waveenergy.html. For more information about EPRI's tidal energy resource assessment see http://oceanenergy.epri.com/streamenergy.html.

[69] On July 6, 2011 DOE released a database, developed in partnership with the Georgia Institute of Technology, of tidal energy resources in the United States. The interactive database is available online at http://www.tidalstreampower.gatech.edu/.

[70] G. Miller, J. Franceschi, W. Lese, and J. Rico, "The Allocation of Kinetic Hydro Energy Conversion Systems (KHECS) in USA Drainage Basins: Regional Resource Potential and Power," New York University, Department of Applied Science, August, 1986.

[71] Capacity factor estimates for hydrokinetic devices were based on data reported by Argonne National Laboratory. See http://teeic.anl.gov/er/hydrokinetic/restech/scale/index.cfm.

[72] As of June 9, 2011, the Federal Energy Regulatory Commission (FERC) had issued 70 preliminary permits for tidal, wave, and inland hydrokinetic projects. For more information, see http://www.ferc.gov/industries/hydropower/indusact/hydrokinetics.asp.

[73] Remarks by Sean O'Neill, President—Ocean Renewable Energy Coalition, at the 14th Annual Congressional Renewable Energy & Energy Efficiency EXPO + Forum, June 16, 2011. DOE maintains an on-line database of ocean and hydrokinetic projects worldwide. For more information, see http://www1.eere.energy.gov/windandhydro/hydrokinetic/default.aspx.

[74] Challenges associated with calculating LCOE for ocean and hydrokinetic electricity generation technologies include (1) unknown capital costs, (2) unknown operations and maintenance costs, (3) unknown technology performance characteristics, etc.

[75] For more information on biomass feedstock, see CRS Report R41440, *Biomass Feedstocks for Biopower: Background and Selected Issues*, by Kelsi Bracmort.

[76] Biomass resource management practices may include land utilization intensity, fertilization, using more productive and/or genetically modified crops, among others. NREL's "Billion Ton" study makes some assumptions for resource management changes needed in order to achieve that resource level. For more information, see "Biomass as Feedstock for a Bioenergy and Bioproducts Industry: The Technical Feasibility of a Billion-Ton Annual Supply," U.S. Department of Energy and U.S. Department of Agriculture, April 2005, available at http://www1.eere.energy.gov/biomass/pdfs/final_billionton_vision_report2.pdf.

[77] Energy Information Administration, see http://www.eia.gov/totalenergy/ data/monthly/pdf/ sec7_5.pdf.

[78] "Biomass as Feedstock for a Bioenergy And Bioproducts Industry: The Technical Feasibility of a Billion-Ton Annual Supply," U.S. Department of Energy and U.S. Department of Agriculture, April 2005, available at http://www1.eere.energy.gov/biomass/ pdfs/final _billionton_vision_report2.pdf.

[79] Biomass Power Association, see http://www.usabiomass.org/. Biomass material might also be co-fired with coal in conventional coal electricity generation facilities. Biomass cofiring with coal may result in improved biomass conversion efficiencies when compared to combusting only biomass. Co-firing biomass with coal may create some technical operating issues associated with tar production and fouling of electricity generating equipment. The degree to which these technical problems might be realized is dependent on the quality of the biomass material being combusted and the percentage of biomass blended and co-fired with coal. For more information see http://www.iea.org/techno/essentials3.pdf.

[80] R. Bain, W. Amos, M. Downing, and R. Perlack, "Highlights of Biopower Technical Assessment: State of the Industry and the Technology," National Renewable Energy Laboratory and Oak Ridge National Laboratory, April 2003, available at http://www. nrel.gov/docs/fy03osti/33502.pdf.

[81] Ibid.

[82] For more information about biomass feedstock characteristics see http://www1.eere.energy. gov/biomass/feedstock_databases.html. For more information about biomass feedstock logistics see http://www1.eere.energy.gov/biomass/feedstocks_logistics.html.

[83] See, for example, National Academy of Sciences, National Research Council, *Electricity from Renewable Resources: Status, Prospects, and Impediments*, National Academies Press, 2010.

[84] Terms such as LCOE, Power Purchase Agreement (PPA), contract price, and others, are sometimes used when discussing renewable electricity economics. Each of these terms has different, sometimes multiple, definitions. For example, the Federal Energy Regulatory Commission (FERC) publishes contract prices for electricity, including renewable electricity projects. Published contract prices for wind generated electricity can range between $40/MWh and $60/MWh. Comparing these contract prices with EIA LCOE estimates ($82/MWh minimum) indicates that wind electricity is being sold for less than cost. However, FERC published contract prices may not reflect any value that the wind project might receive by selling renewable energy credits (RECs). It is important to understand what is being reflected in LCOE, PPA, and contract price values.

[85] For a detailed description of this LCOE methodology, see "The Drivers of Levelized Cost of Energy for Utility-Scale Photovoltaics," SunPower Corporation, August 14, 2008, available at http://nl.sunpowercorp.be/downloads/SunPower_levelized_cost_of_electricity.pdf.

[86] For a description of the cash flow LCOE methodology, see P. Schwabe, S. Lensink, and M. Hand, "IEA Wind Task 26: Multi-National Case Study of the Financial Cost of Wind Energy," IEA Wind, March 2011, available at http://www.ieawind.org /IndexPage POSTINGS/IEA%20WIND%20TASK%2026%20FULL%20REPORT%20FINAL%203%2 010%2011.pdf.

[87] NREL has calculated LCOE estimates for wind and has summarized the sensitivity of LCOE values based on different assumptions. For more information, see K. Cory and P. Schwabe, "Wind Levelized Cost of Energy: A Comparison of Technical and Financing Input Variables," National Renewable Energy Laboratory, October 2009.

[88] For a description of EIA's LCOE methodology and assumptions used for the estimates, see "Levelized Cost of New Generation Resources in the Annual Energy Outlook 2011," Energy Information Administration, December 2010, available at http://www.eia.gov/oiaf/aeo/electricity_generation.html.

[89] Union of Concerned Scientists, 2002, The Hidden Cost of Fossil Fuels, available at http://www.ucsusa.org/clean_energy/technology_and_impacts/impacts/the-hidden-cost-of-fossil.html.

[90] One study from the Brookings Institution attempts to quantify social costs, per unit of energy produced, associated with energy production. For more information, see M. Greenstone and A. Looney, "A Strategy for America's Energy Future: Illuminating Energy's Full Costs," Brookings Institution, The Hamilton Project, May 2011, available at http://www.brookings.edu/~/media/Files/rc/papers/2011/05_energy_greenstone_looney/05_energy_greenstone_looney.pdf. 91 For more information, see "Social Cost of Carbon for Regulatory Impact Analysis Under Executive Order 12866," DOE, Office of Energy Efficiency and Renewable Energy, available at http://www1.eere.energy.gov/ buildings/ appliance_standards/ comercial/pdfs/sem_finalrule_appendix15a.pdf.

[92] The term "power system," for the purpose of this discussion, includes electricity generators, transmission infrastructure, and electricity consumers. For more information about the North American power system, see http://www.nerc.com/page.php?cid=1|15.

[93] "Harnessing Variable Renewables: A Guide to the Balancing Challenge," International Energy Agency, 2011.

[94] Ibid.

[95] North America has three distinct interconnections: (1) Eastern Interconnect, (2) Western Interconnect, and (3) ERCOT (Electric Reliability Council of Texas) Interconnect. Each interconnect essentially operates as an independent electrical grid system.

[96] "Eastern Wind Integration and Transmission Study," National Renewable Energy Laboratory, prepared by EnerNex Corporation, February 2011, available at http://www.nrel.gov /wind/ systemsintegration/pdfs/2010/ewits_final_report.pdf.

[97] "Western Wind and Solar Integration Study," National Renewable Energy Laboratory, prepared by GE Energy, May 2010, available at http://www.nrel.gov/wind/ systems integration/pdfs/2010/wwsis_final_report.pdf.

[98] U.S. Department of Energy, Energy Storage Program Planning Document, http://www.oe.Energy.gov/DocumentsandMedia/OE_Energy_Storage_Program_Plan_Feburary_ 2011 v3. pdf.

[99] Uma Outka, The Renewable Energy Footprint, Stanford Environmental Law Journal, Vol. 30, p. 241, 2011.

[100] Ibid.

[101] On July 21, 2011, the Federal Energy Regulatory Commission (FERC) issued Order No. 1000—Final Rule on Transmission Planning and Cost Allocation by Transmission Owning and Operating Public Utilities. This FERC order may result in transmission capacity access for renewables, since transmission planning must take into account federal and state public policy requirements (i.e., renewable portfolio standards, etc.).

[102] "Eastern Wind Integration and Transmission Study," National Renewable Energy Laboratory, February 2011, available at http://www.nrel.gov/wind/systemsintegration/ pdfs/2010/ewits_final_report.pdf.

[103] "20% Wind Energy by 2030: Increasing Wind Energy's Contribution to U.S. Electricity Supply," U.S. Department of Energy, Energy Efficiency and Renewable Energy, July 2008, available at http://www.nrel. gov/docs/fy08osti/41869.pdf.

[104] National Academy of Sciences, National Research Council, *Electricity from Renewable Resources: Status, Prospects, and Impediments*, National Academies Press, 2010.

[105] NIMBY = Not In My Back Yard.

[106] For more information about potential water issues associated with CSP electricity generation, see CRS Report R40631, *Water Issues of Concentrating Solar Power (CSP) Electricity in the U.S. Southwest*, by Nicole T. Carter and Richard J. Campbell.

[107] "Electricity from Renewable Source: Status, Prospects, and Impediments," Chapter 5 – Environmental Impacts of Renewable Energy, National Academy of Sciences, 2010.

[108] "A National Offshore Wind Strategy: Creating an Offshore Wind Energy Industry in the United States." U.S. Department of Energy, Energy Efficiency and Renewable Energy, February 2011, available at http://www1.eere.energy.gov/windandhydro/ pdfs/national_ offshore_wind_strategy.pdf.

[109] For more information on the commercialization "valley-of-death," see "Crossing the Valley of Death: Solutions to the next generation clean energy project financing gap," *Bloomberg New Energy Finance*, June 21, 2010.

[110] For more information on energy tax policy see CRS Report R41769, *Energy Tax Policy: Issues in the 112th Congress*, by Molly F. Sherlock and Margot L. Crandall-Hollick. For more information on tax favored finance options see CRS Report R41573, *Tax-Favored Financing for Renewable Energy Resources and Energy Efficiency*, by Molly F. Sherlock and Steven Maguire.

[111] For more information see CRS Report R40913, *Renewable Energy and Energy Efficiency Incentives: A Summary of Federal Programs*, by Lynn J. Cunningham and Beth A. Roberts.

[112] For more information regarding Section 1603 of ARRA see CRS Report R41635, *ARRA Section 1603 Grants in Lieu of Tax Credits for Renewable Energy: Overview, Analysis, and Policy Options*, by Phillip Brown and Molly F. Sherlock.

[113] For more information on state incentives for renewable energy, see the Database of State Incentives for Renewables and Efficiency at http://www.dsireusa.org/.

[114] For more information see CRS Report R41720, *Clean Energy Standard: Design Elements, State Baseline Compliance and Policy Considerations*, by Phillip Brown.

[115] One example of this scenario might be the expiration of production tax credits in 2000, 2002, and 2004. For more information, see http://www.awea.org/issues/ federal_policy/ upload/ PTC_April-2011.pdf.

[116] One concept worth noting here is known as the Jevons Paradox, which indicates that as efficiency increases the amount of resources demanded will also increase, not decrease as might be expected. William Jevons, in 1865, observed that as technology improved the efficiency of coal use, consumption of coal actually increased across several industries. Whether or not the Jevons Paradox is applicable today is debatable, with experts presenting arguments that support and refute the Jevons Paradox. A high level overview of the Jevons Paradox is available at http://en.wikipedia.org/wiki/Jevons_paradox. For more information, see CRS Report RL31188, *Energy Efficiency and the Rebound Effect*, by Frank Gottron.

[117] Energy Information Administration, http://www.eia.gov/cfapps/ipdbproject/ IEDIndex3.cfm? tid=92&pid=46&aid=2.

[118] Rich Brown, Sam Borgeson, Jon Koomey, Peter Biermayer, "U.S. Building-Sector Energy Efficiency Potential", Environmental Energy Technologies Division, Ernest Orlando Lawrence Berkeley National Laboratory, Report LBNL-1096E, September 2008.

[119] U.S. Department of Energy, Office of Energy Efficiency and Renewable Energy, http://www.energy.gov/energyefficiency/index.htm, and the U.S. Environmental Protection Agency, http://www.energystar.gov/.

[120] National Renewable Energy Laboratory, http://www.nrel.gov/learning/re_biofuels.html.

[121] Energy Information Administration, http://www.eia.gov/renewable/alternative_transport_vehicles/pdf/afvatf2009. Pdf

[122] U.S. energy use at the national scale is measured in quadrillion British thermal units (Btu), or "quads."

[123] REN21. 2011. Renewables 2011 Global Status Report (Paris: REN21 Secretariat), http://www.ren21.net/REN21Activities/Publications/GlobalStatusReport/GSR2011/tabid/56142/Default.aspx.

[124] Energy Information Administration, Annual Energy Outlook 2011, http://www.eia.gov/ forecasts/aeo/pdf/0383(2011).pdf.

[125] Eckhart, M. "Renewable Energy Exceeds 50 GW and Enters Decade of Scale-Up," Infrastructure Solutions Magazine, April 2011, available at http://www.acore.org/wp-content/uploads/2011/04/Infrastructure-Magazine-Article-V4.pdf.

[126] Energy Information Administration, "Annual Energy Outlook 2011: with projections to 2035," DOE/EIA-0383(2011), April 2011, available at http://www.eia.gov/forecasts/aeo/pdf/0383(2011).pdf.

In: Renewable Electricity Generation   ISBN: 978-1-61942-678-8
Editors: L. G. Irwin and R. Macias   © 2012 Nova Science Publishers, Inc.

Chapter 2

THE EFFECTS OF RENEWABLE OR CLEAN ELECTRICITY STANDARDS[*]

Congressional Budget Office

NOTES

Unless otherwise indicated, all years referred to in this report are calendar years. Numbers in the text, tables, and figures may not add up to totals because of rounding. The cover shows (clockwise from the top) the cooling towers of a nuclear power plant, a geothermal power station, an array of photovoltaic solar panels, wind turbines, a biomass power plant that burns wood chips, and power lines on a transmission tower. (The last photo is courtesy of the National Renewable Energy Laboratory; the others are copyrighted as indicated on the cover.)

PREFACE

Federal lawmakers have recently considered several policies to alter the mix of fuels used to generate electricity in the United States. Those policies—referred to here as renewable or "clean" electricity standards—would lead to

---

[*] This is an edited, reformatted and augmented version of a Congressional Budget Office publication, dated July 2011.

greater reliance on energy sources that produce few or no emissions of carbon dioxide (CO_2), the most prevalent greenhouse gas contributing to climate change. Renewable electricity standards would require a certain share of the nation's electricity generation (say, 25 percent) to come from renewable sources, such as wind or solar power. Clean electricity standards would require a certain percentage of the nation's electricity generation to come from renewable sources or from nonrenewable sources that reduce or eliminate CO_2 emissions, such as nuclear power, coal-fired plants that capture and store CO_2 emissions, and possibly natural-gas-fired plants. Many renewable and clean electricity standards already exist at the state level. This Congressional Budget Office (CBO) study—prepared at the request of the Chairman and Ranking Member of the Senate Committee on Energy and Natural Resources—examines how a federal renewable or clean electricity standard would change the mix of fuels used for electricity generation, the amount of CO_2 emissions, and the retail price of electricity in different parts of the United States. In particular, the study explores how some proposed features of such standards (such as various preferences, exemptions, and alternative compliance rules) would affect those outcomes and identifies underlying causes of uncertainty about such outcomes. The study also highlights key elements in designing a renewable or clean electricity standard that would help minimize its costs to U.S. households and businesses. In keeping with CBO's mandate to provide objective, impartial analysis, this report makes no recommendations.

The study was prepared by Terry Dinan of CBO's Microeconomic Studies Division and by Brian Prest, formerly of CBO, under the direction of Joseph Kile and David Moore. Justin Falk, Daniel Frisk, Ron Gecan, Robert Shackleton, and Julie Somers provided valuable input, and Marin Randall fact-checked the report. Helpful comments also came from Christopher Namovicz of the Energy Information Administration; Patrick Sullivan of the National Renewable Energy Laboratory; David McLaughlin, Karen Palmer, and Matthew Woerman of Resources for the Future; and Catherine Wolfram of the University of California at Berkeley. (The assistance of external reviewers implies no responsibility for the final product, which rests solely with CBO.) Chris Howlett edited the study, with assistance from Sherry Snyder, and John Skeen proofread it. Maureen Costantino and Jeanine Rees prepared the report for publication, and Maureen Costantino designed the cover. Monte Ruffin printed the initial copies, and Linda Schimmel handled the print distribution.

Douglas W. Elmendorf, Director
July 2011

SUMMARY

Many policymakers have expressed interest in mandating that a minimum percentage of the electricity consumed in the United States be generated from renewable or "clean" sources of energy. A majority of states have implemented similar requirements in their jurisdictions. Such requirements— known as renewable or clean electricity standards—would reduce emissions of carbon dioxide (CO_2), the most prevalent greenhouse gas, by decreasing the percentage of electricity generated from fossil fuels. That change would not significantly reduce energy imports, however, because most of the energy used for electricity generation in the United States already comes from domestic sources.

How a Renewable or Clean Electricity Standard Would Work

Currently, only about 10 percent of U.S. electricity is produced from renewable sources of energy, such as hydropower, wind, and biomass (which includes waste products from the forest industry and farms). The bulk of electricity is produced using coal (45 percent), natural gas (24 percent), and nuclear power (19 percent).

Meeting a renewable electricity standard (RES) would generally entail replacing fossil-fuel-fired generation, which emits CO_2, with generation from renewable sources that would produce fewer, if any, CO_2 emissions. In particular, an RES would probably increase reliance on wind, biomass, solar energy, and geothermal energy to generate electricity. Hydroelectric power is usually excluded from RES proposals because of environmental concerns, although it accounts for more than half of all renewable generation at present.

A clean electricity standard (CES) would expand the set of qualifying sources to include not only renewable energy but also nuclear power, which produces no CO_2 emissions, and fossil-fuel-based generation that involves the capture and storage of CO_2 emissions (a process still under development). A more inclusive CES could allow the standard to be met with generation from natural-gas-fired plants, which release only about half as much CO_2 per unit of electricity produced as coal-fired plants do.

Utilities would typically be required to comply with a renewable or clean electricity standard by submitting "credits," each of which certified that a megawatt hour (MWh) of electricity had been produced from a qualifying renewable or other clean source. The number of credits that a utility would have to submit would depend on the standard and on the utility's electricity sales.

For example, under a 30 percent RES or CES, a utility would have to submit 30 credits for each 100 MWhs of electricity it sold.

The federal government would give credits to generators that produced electricity from qualifying sources, and the generators in turn could sell the credits to the highest bidder. Utilities that generate at least some of their own electricity would comply with the policy either by using credits that they received for producing electricity from qualifying sources or by buying credits from other generators that use qualifying sources. Utilities that do not own generating facilities would need to purchase all of their credits. Utilities' demand for credits to comply with the standard would encourage generators to produce more electricity from qualifying renewable or other clean sources. If the credits were traded freely, the market could determine the least expensive method of achieving the desired increase in renewable or clean electricity generation.

Potential Effects on Power Generation, CO_2 Emissions and Electricity Prices

A national RES or CES would alter the mix of energy sources used to produce electricity, the amount of CO_2 emitted, and the price of electricity, with those effects varying by region. To illustrate the effects, the Congressional Budget Office compared the results of seven analyses of different potential federal standards conducted in the past two years by the Energy Information Administration or the National Renewable Energy Laboratory (parts of the Department of Energy) or by the independent research organization Resources for the Future. Those analyses examined renewable and clean electricity standards with a variety of design features and relied on models of the electricity sector that incorporated different assumptions about the costs of relevant technologies. The comparison reveals some common findings about the potential impact of a national RES or CES policy, offers insights into the effects of specific design features, and highlights the uncertainties underlying projections of policy outcomes.

Most analyses concluded that the bulk of the increase in renewable generation resulting from an RES or CES would come from additional wind generation (mainly in the High Plains region of the western and central United States) and from biomass generation (mainly in the Southeast). The relative importance of those sources depends heavily on assumptions about the availability of resources in different regions and about the relative costs of various technologies.

Including certain design features in an RES or CES policy could cause the actual percentage of electricity produced from qualifying sources to be less than the standard. That could happen if some utilities were exempt from complying with the standard; if some technologies were given preferential treatment, allowing them to earn more than one credit per unit of electricity produced; or if utilities were allowed to make "alternative compliance payments" instead of submitting the necessary credits. Those features were either included in RES and CES policies proposed in the previous Congress or are part of some state programs.

Either an RES or CES would reduce CO_2 emissions in the United States compared with the amount that would occur in the absence of the policy. The actual reduction resulting from a given standard and set of design features would be uncertain, however. For example, generators that substituted biomass for coal would reduce emissions more than generators that substituted wind for natural gas, because a MWh of electricity generated from coal produces about twice as much CO_2 as one generated from natural gas.

Either an RES or CES would also raise the average cost of generating electricity in the United States because, in the absence of the standard, regulators and generators would generally choose the lowest-cost method of producing electricity. Higher generation costs in turn would lead to higher electricity prices for many businesses and households; however, the price effects would differ among regions. A federal electricity standard would cause prices to go up in most parts of the country but down in other parts. Predictions about effects on regional electricity prices vary significantly among policies and when different models are used to analyze similar policies. Those effects are strongly influenced by regional patterns of investment in new generating capacity and by the extent to which electricity prices in a given area are set by regulators or determined by market forces.

Changes in electricity prices offer an indication of the effects of an RES or CES on electricity consumers, but they do not provide a comprehensive measure of the policy's overall cost. To the extent that a standard reduced electricity prices in a particular region, the cost of the policy would be borne initially by electricity producers in that region, or by consumers in other regions where utilities (taken together) were net buyers of credits. The cost to electricity producers would take the form of lower returns on their existing capital. Those lower returns would discourage new capital from being invested in the electricity sector, eventually reducing the supply of electricity and causing the price to rise. Thus, ultimately, the cost of the policy would be borne by electricity customers.

Implementing a federal RES or CES would be complicated by the fact that 31 states and the District of Columbia have some form of renewable or clean electricity standard already in place. The incremental effect that a federal standard would have on the amount of renewable or clean generation would depend on the provisions of those state programs. If utilities could not count a given MWh of qualifying generation toward their compliance with both a state and a federal policy, the increase in renewable or clean generation necessary to meet the federal standard—and the cost of achieving that increase—would be much greater than would otherwise be the case. Moreover, regardless of whether a MWh of generation could qualify for credits at both the state and federal levels, the enactment of a federal standard would affect the prices of credits traded in state programs.

As a general rule, a given increase in renewable or clean generation, or a given decrease in emissions, could be accomplished at a lower cost through a single federal standard than through a combination of a federal standard and numerous state standards. The reason is that state policies would tend to constrain the pattern of renewable and clean generation across the United States, hindering the ability of a federal standard to spur the lowest-cost investments in such generation, at least in some regions.

Ways to Make an Electricity Standard More Cost-Effective

Although the costs of meeting a particular RES or CES cannot be predicted with certainty, they could be reduced by incorporating certain design features. For example, allowing unrestricted trading of credits, expanding the range of energy sources that could be used to comply with the policy, phasing in the standard gradually, and giving companies the flexibility to shift credits between years would all make an RES or CES policy more cost-effective.

Unrestricted Trading
The electricity market faces various regional limitations. For example, storing electricity or building transmission lines to move power over long distances is expensive and difficult, and some areas are better suited than others to certain types of generation.

Letting utilities comply with a standard by submitting credits that could be bought and sold independently of the electricity generation with which they were associated—rather than requiring that each utility get a certain percentage of its electricity directly from renewable or other clean sources—would help overcome such limitations and thereby lower utilities' compliance costs.

Compliance costs could be reduced further by allowing financial firms that do not generate or distribute electricity to participate in credit trading. Participation by those firms would increase the liquidity of the market, meaning that utilities and generators could buy and sell large numbers of credits without affecting the price.

Expanded Compliance Options

Allowing as many energy sources as possible to qualify for credits (within the constraints of achieving the objectives of the policy, which might include reducing CO_2 emissions, avoiding further damage to the environment, or developing specific technologies) would help minimize the cost of meeting an RES or CES and of achieving any resulting emission reductions. In particular, a clean electricity standard would be likely to bring about a given reduction in CO_2 emissions at a lower cost than a renewable electricity standard because a CES provides incentives for a wider variety of low-emitting technologies than an RES does.

If regulators linked the amount of credits that various technologies could receive to their emissions, then letting both existing and new sources of electricity generation earn credits (rather than just sources that started operating after the policy began) could help better align financial incentives with actual emission reductions.

For example, granting partial credits for both existing and new natural-gas-fired generation would give generators a larger financial incentive to substitute a megawatt hour of emission-free generation for a megawatt hour of generation from a high-emitting source, such as coal, than from a low-emitting source, such as natural gas.

Total costs of reducing CO_2 emissions could be lowered even further by allowing emission-reducing improvements in energy efficiency to qualify for credits. For example, generators could upgrade their plants in a manner that allowed them to produce the same amount of electricity from less fossil fuel, or large companies could install lighting that used less electricity.

However, regulators would face significant challenges in accurately measuring the energy or fuel savings from such improvements.

Even with a wide variety of compliance options, neither an RES nor a CES would be as cost-effective in cutting CO_2 emissions as a "cap-and-trade" program. Such a program would involve setting an overall cap on emissions and letting large sellers of emission-creating products (such as electricity generators, oil producers and importers, and natural gas processors) trade rights to those limited emissions.

In that way, a cap-and-trade program would create a direct incentive to cut emissions; in contrast, an RES or CES would create a direct incentive to use more renewable or other types of clean electricity but would have only an indirect effect on emissions.

Gradual and Flexible Timing

Electricity generation typically involves investments in large-scale and long-lasting physical equipment, and U.S. demand for electricity is growing slowly enough that the potential for investment in new generation and distribution capacity is fairly small. Utilities and generators would therefore benefit from provisions that phased in an RES or CES gradually over an extended period.

They would also benefit from being allowed to transfer credits between different time periods—by "banking" current excess credits for use in later years or by "borrowing" credits that they expected to earn in the future for use now. Such provisions would make it easier for utilities and generators to comply with the standard in the course of planning for moderate increases in new capacity, without prematurely retiring existing capacity. However, the standard would not be met if firms that borrowed credits failed to fulfill their obligations.

1. INTRODUCTION

In recent years, federal policymakers have proposed various ways to alter the mix of energy sources used to generate electricity in the United States. In many cases, those proposals have been motivated by a desire to decrease emissions of carbon dioxide (CO_2) that result from burning fossil fuels. CO_2 is the most prevalent of the greenhouse gases, whose accumulation in the atmosphere, if allowed to continue unabated, is expected by a strong consensus of experts to have extensive, highly uncertain, but potentially serious and costly impacts on regional climates throughout the world.

Some of those proposals involve renewable electricity standards (RESs), which would require utilities to meet at least a certain percentage of consumers' demand for electricity with power generated from renewable sources, such as wind or solar energy. Other proposals involve clean electricity standards (CESs), which would require utilities to use some minimum percentage of "clean" energy sources—generally defined to include renewable sources and also nuclear power (which emits no CO_2) and fossil-fuel-burning power plants that capture and compress their CO_2 emissions for storage underground (a technology that is still at the development stage). Some CES proposals would also define natural gas as a clean energy source because it emits roughly half as much CO_2 as coal does per unit of electricity produced. RESs and CESs in some form are already operating in a majority of U.S. states.

Under a federal renewable or clean electricity standard— as under many of the state policies—electric utilities would have to comply with the standard by submitting "credits" to the government. Each credit would represent a megawatt hour (MWh) of electricity that was generated from a qualifying renewable or other clean source. For example, a 30 percent RES or CES would obligate utilities to submit 30 credits for each 100 MWhs of electricity they sold. In general, utilities would buy the credits they needed from electricity generators, who would receive credits from the government for each MWh of qualifying electricity they produced. The demand for credits on the part of utilities would encourage generators to produce more electricity that qualified for credits. If the credits could be traded freely, the market could determine the least costly method of achieving the desired increase in renewable or clean electricity generation.

This chapter provides a brief overview of federal proposals and existing state standards for renewable and clean electricity, describes how a federal standard would work, examines the current mix of energy sources used to produce electricity, and discusses the challenges to increasing the use of renewable or other clean sources.

Chapter 2 assesses how a federal RES or CES might affect power generation and electricity prices in various regions of the United States and the amount of CO_2 emitted by the electricity sector. It also discusses complications posed by layering a federal standard on top of existing state standards. Chapter 3 offers general observations about how an RES or CES policy could be designed to be cost-effective.

Federal Proposals and State Standards for Renewable and Clean Electricity

During the previous Congress (the 111th), several Senators proposed legislation to establish renewable or clean electricity standards. For example, Senators Jeff Bingaman and Sam Brownback proposed an RES in the Renewable Electricity Promotion Act of 2010 (S. 3813).

Under that standard, 15 percent of U.S. electricity generation would have had to come from renewable sources by 2021. Senators Richard Lugar, Lindsey Graham, and Lisa Murkowski introduced the Practical Energy and Climate Plan Act of 2010 (S. 3464), which would have required that, by 2030, 30 percent of electricity generation come from renewable sources, from new nuclear capacity, or from coal plants that capture and store CO_2 emissions. Both of those proposals would have utilities meet part of their compliance obligation by submitting energy-efficiency credits. Such credits could be awarded to generators that upgraded their facilities in a manner that allowed them to generate more electricity with the same amount of fossil fuel, to states or utilities that sponsored programs to promote energy-saving improvements, or to large end users of electricity that made energy-saving investments. (Neither bill was brought to a vote, and neither has been reintroduced in the 112th Congress.)

In addition, in his 2011 State of the Union address, President Obama expressed support for a CES that would require 80 percent of electricity generation to come from clean sources by 2035. Under that proposal, clean sources would include renewable energy, nuclear power, "efficient" natural gas, and fossil-fuel-fired plants that capture, compress, and store their CO_2 emissions.

In recent decades, 31 states have implemented renewable or clean electricity standards of varying stringencies:

27 states plus the District of Columbia have renewable portfolio standards, which are similar to renewable electricity standards, and another 4 states have alternative energy portfolio standards, which are similar to clean electricity standards (see Figure 1-1).[1] In addition, 7 states have voluntary goals for electricity generation. (The other 12 states have no mandatory standards or voluntary goals.)

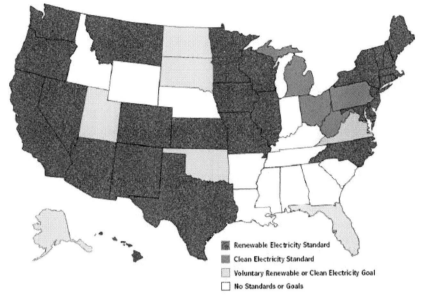

Source: Congressional Budget Office based on information from the Pew Center on
    Global Climate Change.
Note: States vary widely in which sources of electricity generation are eligible to
    receive credits under their standards. For example, West Virginia provides credits
    for certain types of coal technology, although most other states do not.

Figure 1-1. States with Renewable or Clean Electricity Standards or Goals.

How a Renewable or Clean Electricity Standard Could Work

Virtually any renewable or clean electricity standard would have some
basic features, such as the standard itself (the minimum share of electricity
generated by qualifying sources that the policy would require), definitions of
the energy sources included in the policy, and credits that utilities would
submit to comply. Typically, the standard would be phased in over time; for
example, a CES might require that 50 percent of electricity be generated from
clean sources by 2025, with lower but increasingly strict standards for each
year leading up to 2025.

An RES or CES could have a variety of other design features as well, such
as allowing credits to be traded widely, varying the number of credits that
sources are awarded according to their CO_2 emissions, providing credits to
existing generators (rather than limiting them to new generating capacity),

allowing credits to be "banked" or "borrowed" for use in other years, and let-ting improvements in energy efficiency qualify for credits. Moreover, policymakers could opt to limit the cost of complying with the standard by letting utilities make alternative compliance payments instead of submitting credits, could give preference to certain energy technologies, and could combine an RES and a CES into a tiered standard. Those various design features, which are described below, would influence a policy's effects on electricity production, electricity prices, and CO_2 emissions (as discussed in Chapter 2) and the policy's cost-effectiveness (as discussed in Chapter 3).

Credits Could Be Traded Widely

Credits would be tradable financial assets that could be bought and sold independently of the electricity generation with which they were associated. Generators would receive credits from the federal government when they produced electricity from qualifying sources and could sell those credits to utilities, either with or without the associated power. Utilities would use the credits to comply with the policy. Utilities that generate at least some of the electricity they sell would comply either by using credits that they received for producing electricity from renewable sources or buying credits from other generators that use qualifying sources. Utilities that do not own generating facilities would have to buy all of their credits.

The U.S. electricity market tends to be regional in scope. Once generated, electricity is difficult and costly to store, and moving it from one place to another requires the use of transmission lines, which are expensive and time consuming to build. In addition, as discussed later in this chapter, some areas are better suited than others to specific types of power generation. Allowing utilities to comply with an RES or CES by submitting credits would get around such constraints by decoupling utilities' compliance from the sources of energy used to generate the electricity they sold. Thus, for example, a utility in Ohio could comply with the standard by purchasing credits from a wind plant in North Dakota, without actually purchasing the plant's electricity (which, for logistical reasons, would probably be sold to a local utility).

As the market for credits developed, entities other than generators and utilities—such as banks and pension funds—could be allowed to trade credits. Participation by those entities would provide liquidity to the market, enabling utilities and generators to buy and sell large numbers of credits without affecting the price. Utilities would seek to purchase credits at the lowest possible price, and that competitive pressure would encourage generators that could expand their electricity production from qualifying sources at the lowest cost to do so. The price of credits would rise to the point at which the number of credits created by qualifying generators was large enough for utilities to obtain the credits they needed to comply with the standard.

Credit Amounts Could Be Related to CO_2 Emissions

The amount of credits that generators would receive for various types of electricity production could be adjusted to account for differences in the amount of CO_2 released. For example, a megawatt hour of electricity generated at a nuclear plant would produce no CO_2 emissions and thus would receive one credit under a CES policy. A megawatt hour generated at a natural-gas-fired plant would emit roughly half as much CO_2 as the same amount of generation at a coal-fired plant (in the absence of capture and storage technology) and thus might receive half a credit per MWh. Such weighting would give generators the greatest incentive to install the cleanest technologies and, as a result, would help achieve emission reductions in a cost-effective manner. However, it would require policymakers to define partial credits for all technologies in proportion to their emissions and update those definitions over time to reflect technological changes.

Credits Could Be Given to Existing Generators

An RES or CES policy could allow both existing and new sources of electricity generation to earn credits, or it could restrict eligibility to sources that began operating after the program started. In order to achieve the same amount of renewable or clean generation, the stringency of the standard would need to be adjusted accordingly. For example, nuclear power, hydropower, and other renewable sources currently account for about 30 percent of U.S. electricity generation; until the existing stock of power plants was replaced, a 40 percent CES in which existing generators of electricity from those sources were eligible to receive credits would lead to the same increase in clean electricity as a 10 percent CES in which credits were not granted to existing generators.

The decision about whether to allow existing sources of electricity generation to earn credits would have consequences for how the policy's effects were distributed as well as for its ability to motivate the lowest-cost emission reductions:

- If existing sources could receive credits—and thus more credits would have to be required and purchased to produce the same increase in qualifying generation—the additional purchases would transfer revenue from areas that were net buyers of credits to areas that were net sellers. In general, regions with existing plants that generate qualifying electricity—such as the Southeast, which has a significant amount of existing nuclear generation—would benefit from the sale of credits. At the same time, utilities (and ultimately their customers) in other regions would face higher costs because of the need to buy a larger number of credits.

- Granting credits to existing sources would also motivate lower-cost emission reductions than would otherwise be the case because the policy would provide incentives to retire relatively dirty sources of electricity generation earlier than relatively clean sources. For example, plant operators would be more likely to retire coal-fired plants first if existing nuclear or gas-fired plants received credits for the electricity they produced.

Credits Could Be Banked and Borrowed

An RES or CES program could allow utilities to transfer credits between years by letting them bank any excess credits they possessed for use in a future year. The program could also let utilities borrow credits from future years—for example, by permitting them to comply in 2015 with a credit they would obtain in 2016. A utility might want to borrow credits from the following year if it anticipated that a new qualifying plant would begin generating electricity in that year, increasing the supply and lowering the price of credits. As discussed in Chapter 3, banking provisions could help reduce the overall cost of complying with the standard by allowing utilities to stock up on credits in years when they were relatively inexpensive. Most state programs permit banking of credits, but not borrowing.

Improvements in Energy Efficiency Could Qualify for Credits

The types of activities that could earn credits could be expanded to include qualified electricity savings— reductions in electricity consumption resulting from efficiency improvements that would not have been made in the absence of an RES or CES. Such efficiency improvements could occur in the production of electricity; for example, plant operators could decrease the amount of fossil energy needed to produce a MWh of electricity. Efficiency improvements could also occur in the consumption of electricity; for instance, businesses and households could invest in insulation or energy-efficient lighting.

Granting credits for energy-efficiency improvements could, in theory, provide a financial incentive for low-cost reductions in CO_2 emissions. Implementing that feature would be challenging, however. Policymakers would need to clearly define criteria for what types of efficiency improvements might qualify for credits and then determine whether each improvement met the criteria. For instance, establishing the legitimacy of a firm's claim for energy-efficiency credits would require establishing a baseline to determine what the company's energy consumption would have been in the absence of the efficiency improvement. Ideally, that assessment would account for factors that affect the firm's energy use (independently of the efficiency improvements), such as variations in weather or demand for its products.

Compliance Costs Could Be Limited

Policymakers could set an upper limit on the cost of complying with the standard by letting utilities make an alternative compliance payment (ACP) instead of submitting a credit. In that case, utilities would be willing to purchase credits from generators up to the point at which the price of a credit equaled the ACP; beyond that point, they would choose to comply by making the alternative payment. As a result, renewable or clean generation would increase until the legislated standard was met or the additional cost of producing electricity from qualifying sources was equal to the ACP. If utilities complied with a standard by making alternative payments, the actual percentage of electricity coming from renewable or other clean sources would be less than the stated standard.

An ACP could dampen the price of credits and, if utilities were able to bank credits, could do so even when the market price for credits was less than the alternative payment. The reason is that utilities would attach a lower value to a credit today to reflect the fact that its price could not rise above the ACP at any future time.[2] Such price dampening would be most likely to occur when the market price of credits was near the ACP. Any reduction in the price of credits that resulted from the ACP would reduce the incentives for utilities to invest in renewable or other clean technologies.

If policymakers chose to include an ACP, they would need to decide how to allocate any revenue it raised. Those choices could have a variety of implications for the cost-effectiveness of the program and for the manner in which its costs would be distributed among regions or households.[3] Policymakers could use the revenue collected from the ACP to offset costs for some of the producers and consumers who would be affected by the standard.

Certain Technologies Could Receive Preferential Treatment

Policymakers could also design an RES or CES to favor selected technologies—either by establishing different standards for different technologies (referred to as tiered standards) or by providing extra credits for certain technologies. Several states' standards include provisions that provide such preferential treatment. For example, Nevada has a tiered standard requiring that, in 2013, 5 percent of electricity generation in the state come from solar power and a total of 15 percent come from all qualifying renewable sources.[4] In another example, Michigan provides triple credits for each MWh of electricity generated by photovoltaic solar power (which converts solar radiation into direct-current electricity using semiconductors).

The reasons for giving preferential treatment to particular technologies vary. In some cases, states want to encourage types of generation for which their state is particularly well suited. In other cases, favorable treatment goes to sources that are expected to have a large potential for cost reductions in the future because of "learning by doing" (the process through which innovations occur as a result of gaining experience with a technology). The potential for learning by doing is generally thought to be higher for fairly new sources of electricity generation, such as photovoltaic power, than for technologies that have been in use for many years, such as onshore wind generation. Advocates of extra incentives for relatively new or rapidly changing technologies suggest that such incentives could spur enough additional innovation to significantly lower the cost of those technologies.

Clean and Renewable Standards Could Be Combined

A CES and an RES could be combined in a tiered fashion. For example, a program could require that, by a certain year, 50 percent of all electricity be generated from clean sources (including renewable energy, nuclear power, and coal- or natural-gas-fired plants with carbon capture and storage) and that at least 25 percent of electricity be provided exclusively from renewable sources. In that case, utilities would need to ensure that at least half of the credits they submitted came from renewable sources. Complying with that tiered standard could cost more than complying with a 50 percent CES. However, advocates of tiered standards suggest that the additional cost could be justified because renewable sources would impose lower environmental and health costs than alternative clean sources. For example, using renewable sources to generate electricity would not involve creating dangerous waste materials, as nuclear generation does, or pollution runoff, as mining coal to use in coal-fired plants with carbon capture and storage would.

TRENDS AND CHALLENGES IN PRODUCING ELECTRICITY FROM RENEWABLE AND OTHER CLEAN SOURCES

An RES or CES would change the mix of sources for electricity generation. In the absence of additional policy initiatives, that mix is projected to remain fairly constant over the next two decades, in part because the expansion of renewable or clean generation is limited by various factors that make those sources relatively costly.

Trends in the Generation Mix

Coal is currently the leading fuel used to produce electricity in the United States, accounting for 45 percent of generation in 2010 (see Figure 1-2). It is followed by natural gas (24 percent), nuclear power (19 percent), petroleum (1 percent), and a variety of renewable sources (totaling 10 percent). According to the Department of Energy's Energy Information Administration (EIA), that mix of sources is unlikely to change much over the next 25 years under current policies. Coal is expected to remain dominant—declining only slightly as a share of total generation, to 43 percent in 2035—because it is abundant, relatively inexpensive, and well suited to the continuous operation of plants that provide "base-load" generation (electricity to meet the minimum level of demand around the clock). Natural gas and nuclear power are also expected to account for roughly the same shares of total generation in 2035 that they do today. Fossil-fuel-burning plants that capture and store CO_2 emissions do not yet exist on a commercial scale and thus do not provide any significant generation; in the absence of additional policies, no significant amount is projected for the next 25 years.

Of the sources of energy that are naturally replenished, hydroelectric power is the main one in use today, accounting for 6 percent of electricity generation. However, many RES programs exclude hydropower as a qualifying renewable source because of various concerns, such as those about the effects of dam construction on stream flow and fish migration and the displacement that occurs when land is flooded to create reservoirs.

Among renewable sources other than hydropower, wind currently accounts for the bulk of electricity generation (see Figure 1-3 on page 11), followed by biomass (materials such as municipal solid waste, residue from crops, and wood waste from the timber industry) and geothermal energy (the

Earth's internal heat).[5] Solar power accounts for only a tiny fraction of electricity generation. EIA projects that biomass will make up a growing share of nonhydropower renewable generation in the future, roughly equaling wind power by 2035, and that most of the biomass will be used by industrial producers that generate electricity to use in producing finished goods.

The total share of U.S. electricity generated from sources with very low emissions of CO_2—renewable energy and nuclear power—was 29 percent in 2010 and is expected to grow only slightly, to 31 percent, by 2035. EIA projects only modest growth in those sources, for reasons discussed in the next section, despite the fact that various federal grants, tax credits, and subsidies are designed to encourage such generation (see Box 1-1).

Challenges Associated with Increasing Renewable or Clean Generation

Increasing the share of electricity generated from renewable sources, nuclear power, or fossil-fuel-fired plants with carbon capture and storage would reduce CO_2 emissions (compared with what they would be otherwise). Achieving such an increase, however, would require overcoming a variety of complications that generally make renewable or clean generation more costly than fossil-fuel-fired generation. Those complications include the location-specific and variable nature of some renewable energy sources, technological uncertainties, and environmental concerns. Substantially expanding the use of low-emitting sources to generate electricity—through an RES, CES, or other policy—would require addressing those factors.

Location Constraints. Most forms of renewable generation are better suited to some locations than others, and many of those places are far from large population centers that have high electricity demand.6 For example, land areas that are well suited to wind generation—places with average annual wind speeds of at least 6.5 meters per second at a height of 80 meters—are concentrated in the necessary to make those enhanced methods cost-competitive middle of the United States (see Figure 1-4 on page 12), whereas the population is concentrated along the coasts.

Relatively modest increases in the use of renewable sources could be made without significantly changing the infrastructure for distributing electricity: Those increases could take place in windy or sunny areas and be used to meet local demand for electricity. However, substantially boosting reliance on renewable generation would require transporting the power to areas that consume large amounts of electricity—a process that would entail costly and time-consuming expansions of transmission capacity.

(Billions of kilowatt hours)

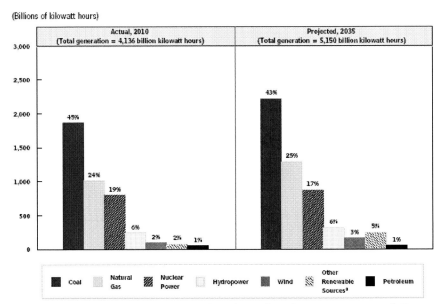

Source: Congressional Budget Office based on data from Department of Energy, Energy Information Administration, Annual Energy Outlook 2011, with Projections to 2035 (April 2011), www.eia.gov/forecasts/aeo/.

a. Includes wood waste products and other biomass burned as fuel, geothermal energy, municipal waste (trash) burned to reduce its volume and to produce electricity, and solar energy.

Figure 1-2. Mix of Sources Used to Generate Electricity in the United States in 2010 and 2035, by Amount of Electricity Provided and Share of Total Generation.

Finding sites for new transmission lines can require obtaining permission from numerous government entities, because many lines cross multiple counties and states. On average, siting a new transmission line takes 14 years.[7] In addition, building transmission lines entails resolving difficult questions about how the cost of that capacity should be allocated.[8] Once constructed, transmission lines could be used by generators or customers who were not envisioned in the initial plan; thus, the generators and customers who agreed to an initial allocation of cost might need to renegotiate that plan as additional generators and customers wanted to make use of the new lines.

Intermittency. Wind and solar plants can be operated only part of the time because the wind does not always blow and the sun is not always visible. On average, wind plants produce just 34 percent of the electricity that they could if they operated continuously; solar plants produce 22 percent to 31 percent of their theoretical full capacity, depending on the type of plant.[9]

Box 1-1. Current Policies Aimed at Lowering the Costs of Renewable and other Clean Electricity Generation

The amount of U.S. electricity generated from renewable sources other than hydropower is small (currently about 4 percent of total generation), despite the fact that such generation has received significant subsidies over the past several decades. For example, renewable electricity production received more than $1 billion in federal subsidies in fiscal year 2007 (the most recent year for which comprehensive measures are available), or 15 percent of all subsidies for electricity generation that year (see the table on page 9). The bulk of the subsidies for renewable electricity production went to wind generation, amounting to $23 per megawatt hour produced, or 2.3 cents per kilowatt hour (kWh). By comparison, the average retail price of electricity sold in the United States in 2007 was 9.1 cents per kWh.[1]

Until 2009, federal support for renewable energy projects primarily took the form of tax credits for production or investment. For instance, qualifying wind generation was eligible for a production tax credit for 10 years after a facility was put in place. Solar generation, by contrast, mainly received support through a 30 percent investment tax credit. Section 1603 of the American Recovery and Reinvestment Act of 2009 (ARRA, Public Law 111-5) created a $5.6 billion program to provide cash grants for renewable energy projects that were placed in service during the 2009 or 2010 tax years or that met certain start-of-construction requirements before the end of 2010. Firms could choose to receive those grants instead of the production or investment tax credits for which they would otherwise be eligible. Projects involving most forms of renewable electricity generation could receive grants equal to 30 percent of their eligible cost basis.[2] The grant program was later extended through 2011 by the Tax Relief, Unemployment Insurance Reauthorization, and Job Creation Act of 2010 (P.L. 111-312).

Wind generation has received about 84 percent of that ARRA grant money.[3] As a result, such generation is on track to more than double over a four-year period, rising from 56 billion kWhs in 2008 to 143 billion kWhs (or about 4 percent of projected U.S. electricity generation) in 2012. However, the Energy Information Administration projects that after the cash grant program expires in 2012, wind generation will remain fairly flat, rising only to 159 billion kWhs by 2030.[4]

Nonrenewable "clean" energy sources have also received major federal support over the years. Significant incentives for building new nuclear power plants were included in the Energy Policy Act of 2005.

Federal Subsidies and Support for Electricity, Fiscal Year 2007

	Net Generation in 2007	Federal Subsidies and Support in 2007	
	(Billions of kilowatt hours)	Millions of 2007 Dollars	Dollars per Megawatt Hour of Electricity Produced
Electricity Generation			
Coal	1,946	854	0.44
Refined coal	72	2,156	29.81
Natural gas and petroleum liquids	919	227	0.25
Nuclear power	794	1,267	1.59
Renewable energy			
Biomass (and biofuels)	40	36	0.89
Geothermal energy	15	14	0.92
Hydropower	258	174	0.67
Solar energy	1	14	24.34
Wind	31	724	23.37
Landfill gas	6	8	1.37
Municipal solid waste	9	1	0.13
Unallocated renewable energy[a]	*	37	*
Subtotal, renewable	360	1,008	2.80
Electricity Transmission and Distribution	n.a.	1,235	n.a.
Total	4,091	6,747	1.65

Source: Congressional Budget Office based on Department of Energy, Energy Information Administration, Federal Financial Interventions and Subsidies in Energy Markets 2007 (April 2008), Table 35, p. 106.

Note: * = reported as too small to be meaningful; n.a. = not applicable.

[a] Includes projects funded under the Clean Renewable Energy Bonds and Renewable Energy Production Incentive programs.

They included production tax credits (1.8 cents per kWh for qualifying new generation), loan guarantees, insurance against regulatory delays, and extension of the Price-Anderson Act, which set a cap on the total amount of liability that the industry could incur and defined rules for apportioning that liability. In fiscal year 2007, nuclear generation received more than $1.2 billion in subsidies, or about 0.16 cents per kWh. Even with those subsidies, future growth in nuclear generation is uncertain.

New federal subsidies and incentives had prompted some renewed interest (a conditional commitment was made for an $8.3 billion guarantee for a loan to finance the addition of two reactors at Southern Company's Plant Vogtle in Georgia). However, concerns about financial risks and health and environmental safety persist, particularly in light of the recent disaster at the Fukushima Daiichi nuclear plant in Japan.

Moreover, in recent years, the Congress has appropriated an average of about $400 million annually to develop technology to reduce carbon dioxide (CO_2) emissions from coal, most notably by capturing and storing CO_2 produced by coal-burning plants and industrial facilities. (That 2007 funding is included in the $854 million for coal shown in the table.)

The Congress also appropriated $3.4 billion in 2009 under ARRA for those activities. In spite of such support, no significant fraction of current electricity generation involves carbon capture and storage, and none is projected to do so for the next two decades.

As a result of those low "capacity factors," the capital costs of building wind and solar plants are spread over a fairly small amount of generation, and hence the plants tend to have relatively high average costs. That intermittency, together with a lack of a low-cost way to store electricity, also means that wind and solar power cannot serve as a source of continuous electric power, as base-load generation through coal combustion does. Finally, periods of high wind or bright sunshine may not correspond with the periods when the value of electricity, and thus the price that generators could charge for it, is highest.[10] That fact tends to limit the potential for renewable generation to be dispatched (or increased quickly) when the demand for electricity is greatest.

In general, wind and solar plants are more frequently substituted for natural-gas-fired generation than for coal-fired generation, because natural gas tends to have higher operating costs than coal and because coal plants tend to operate continuously. However, for the reasons described above, an additional megawatt of solar or wind capacity cannot substitute for an additional megawatt of nonrenewable capacity. For example, given the current generation mix in the United States, EIA estimates that 50 megawatts of wind capacity would be required to replace 20 megawatts of natural-gas-fired capacity. In the absence of a low-cost means of storing electricity, the intermittent nature of wind and solar energy would significantly limit their ability to provide the majority of the nation's electricity, even if all regions were equally well suited for such generation.

(Billions of kilowatt hours)

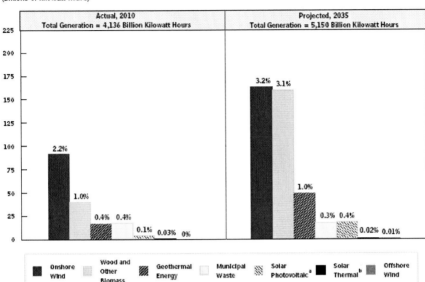

Source: Congressional Budget Office based on data from Department of Energy, Energy Information Administration, Annual Energy Outlook 2011, with Projections to 2035 (April 2011), www.eia.gov/forecasts/aeo/.

a. Solor photovoltaic power involves converting sunlight directly into electricity through the use of special cells.

b. Solar thermal power involves using sunlight to create heat, which can then serve as a power source in a conventional electricitygenerating plant (among other applications).

Figure 1-3. Nonhydropower Renewable Sources of Electricity Generation in 2010 and 2035, by Amount of Electricity Provided and Share of Total Generation.

Potential Effects of Large-Scale Use of Biomass. Unlike wind and solar power, biomass can be used to produce electricity around the clock; therefore, biomass may be able to substitute for coal in supplying continuous base-load generation. However, as reliance on biomass grows—for producing biofuels for the transportation sector as well as for use in the electricity sector—its price may increase. Current biomass sources include waste products generated by the forest products industry and by farms. Increased reliance on biomass to generate electricity and to supply transportation fuels could involve growing crops explicitly for that purpose, which could drive up prices for land and for the agricultural commodities that the biomass crops would displace.[11] One estimate suggests that growing enough biomass to satisfy 6 percent of total

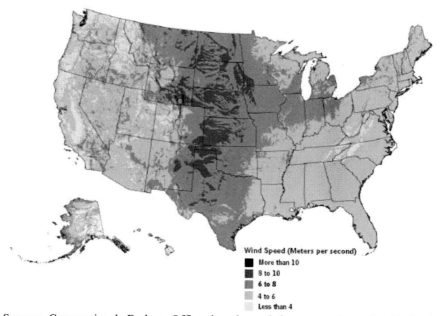

Source: Congressional Budget Office based on information from the National
   Renewable Energy Laboratory, www.windpoweringamerica.gov/wind_maps.asp.
Note: Locations with an average annual wind speed of at least 6.5 meters per second at
   a height of 80 meters are considered well suited to generating wind power.

Figure 1-4. Average Annual Wind Speeds in the United States at 80 Meters Above the
Ground.

U.S. electricity demand in 2008 would have required an amount of land
roughly equal to that currently being farmed for all types of crops in Iowa.[12]

Burning biomass specifically grown for electricity generation in place of
fossil fuels could also raise questions about the extent to which such a
substitution would actually reduce CO_2 emissions. Determining the ultimate
effect of that substitution would require comparing all of the emissions that
result from producing, distributing, and burning both types of fuel (so-called
life-cycle emissions). Such calculations are complicated and depend on
assumptions about whether land-use patterns would be altered by growing
biomass for energy production.[13]

Technological, Safety, and Environmental Uncertainties. No
commercial-scale electricity plants currently exist that capture, compress, and
store CO_2 emissions. As a result, the cost of building such a plant is uncertain,
although both construction and operating costs would be higher than for an
equivalent new plant without that capability. A generating plant designed for

(Percentage of total generating capacity)

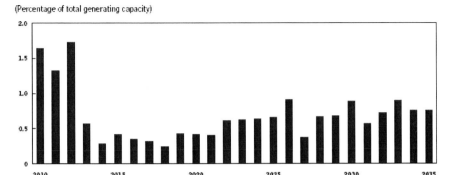

Source: Congressional Budget Office based on data from Department of Energy, Energy Information Administration, Annual Energy Outlook 2011, with Projections to 2035 (April 2011), www.eia.gov/forecasts/aeo/.

Note: Added capacity is relatively high from 2010 through 2012 in part because of cash grants provided through the American Recovery andReinvestment Act of 2009 that are set to expire at the end of 2011.

Figure 1-5. Projected Annual Additions to U.S. Generating Capacity Through 2035.

carbon capture and storage would need additional equipment to separate CO_2 from the rest of the exhaust gases, capture it, and compress it so that it could be transported to storage sites. Moreover, capturing and compressing CO_2 is an energy-intensive process, which would consume 15 percent to 30 percent of the total electricity that the plant would produce. As a result, the plant would need to be larger than one that sold the same amount of electricity but did not capture and store its CO_2 emissions, and it would require more inputs for each MWh of electricity that it sold. Because no full-sized plants with that capability have yet been built, available cost estimates rely on engineering designs. However, as technologies pass through different stages from development to commercialization, researchers often discover that the actual implementation of a concept is more complicated, and hence more expen expensive, than it first appeared.[14]

Expanding carbon capture and storage could also be hampered by environmental or safety concerns about the potential for unintentional releases of compressed CO_2. Likewise, new nuclear power plants would raise safety and environmental concerns. Opponents of nuclear power cite potential risks and environmental damage from mining, processing, and transporting uranium; the risk of nuclear weapons proliferation; the unsolved problem of storing nuclear waste safely; and the potential hazard of a serious accident. Such concerns make developing new nuclear generating capacity difficult, time consuming, and costly. As a result, no new nuclear plant has begun operating in the United States during the past 15 years.

Limited Growth in New Capacity. Altering the generation mix could be achieved at the lowest possible cost through an RES or CES if generators could simply invest in qualifying generation when they needed to add capacity. In that case, the share of generation provided by qualifying sources would increase gradually, without forcing the premature retirement of existing capacity.[15]

However, the demand for new generating capacity in the United States is expected to be relatively low in coming decades. Predictions from EIA suggest that annual construction of new capacity will account for less than 1 percent of total generating capacity each year between 2013 and 2035 (see Figure 1-5), and cumulative new capacity over that period will make up just 12 percent of generating capacity in 2035. Capacity growth is projected to be particularly slow in the near future: Annual construction of new capacity is expected to be less than 0.7 percent of total capacity each year between 2013 and 2025, and cumulative new capacity over that time frame is expected to account for only 6 percent of capacity in 2025.

The limited demand for construction of new generating capacity stems from a variety of factors, including:

- Excess generating capacity in the electricity sector today—driven in part by incentives for new wind generation (see Box 1-1 on page 9)—which lessens the need for firms to replace capacity that is due to be retired in the next decade;

- The fact that virtually no existing capacity is sched- uled to be retired between 2021 and 2031; and

- Slow projected growth in the total demand for electricity in the United States over the next 25 years. The growth rate of electricity production has gradually declined over the past several decades, from an average of 4 percent a year in the 1970s, to 3 percent in the 1980s, 2 percent in the 1990s, and less than 1 percent in the past decade. Much of that slowdown in growth reflects the fact that the U.S. economy has become less electricity intensive per unit of output. For example, between 1980 and 2009, electricity generation grew much more slowly, on average, than real (inflation-adjusted) gross domestic product (by 1.8 percent a year versus 2.8 percent).[16]

- By itself, slow growth in the demand for electricity could reduce the cost of meeting an RES or CES. The reason is that achieving the standard would require a smaller absolute increase in qualifying generation if growth in demand was slow than if it was rapid. However, coupled with slow turnover in the existing capital stock, slow growth in demand for electricity could increase the likelihood that meeting the standard would force the premature retirement of existing power plants.

2. POTENTIAL EFFECTS OF A RENEWABLE OR CLEAN ELECTRICITY STANDARD

A renewable or clean electricity standard would directly alter the mix of sources used to generate electricity in the United States, which in turn would change the amount of greenhouse gases and other pollutants emitted by electricity generators and the prices charged for electricity. Those effects would vary across the country.

In addition, some advocates of a federal RES or CES suggest that it would reduce U.S. dependence on foreign sources of energy. Such an outcome is unlikely, however, because the vast majority of fuel used to generate electricity in the United States—primarily coal and natural gas—already comes from domestic sources. (The United States produces enough coal to be a net exporter, and only about 12 percent of the natural gas it consumes is imported, on net.) Further, because petroleum accounts for just 1 percent of electricity generation, an RES or CES would not significantly affect U.S. oil consumption or imports.[1]

To better understand the potential impact of a renewable or clean electricity standard, the Congressional Budget Office (CBO) examined seven recent analyses of potential federal standards conducted by various government or independent organizations (see Table 2-1).[2] Each policy that was analyzed represents a unique combination of features. Three of the policies would establish a 25 percent RES, but they differ in terms of whether utilities could make an alternative compliance payment (ACP) instead of submitting credits and whether generators would receive multiple credits for certain technologies (those design features are described in Chapter 1). Two of the policies would establish a 25 percent CES, one of which would allow natural gas generation at plants that began operation after the policy took effect to qualify for credits and one of which would not. The other two policies would establish a 20 percent RES, one with an ACP provision and one without. The seven analyses simulated the effects of those policies using three models of the electricity sector:

- The National Energy Modeling System (NEMS), which was developed by the Department of Energy's Energy Information Administration (EIA);
- The Regional Energy Deployment System (ReEDS) model, which was created by the National Renewable Energy Laboratory (NREL), also part of the Department of Energy; and
- Haiku, which was developed by the independent research organization Resources for the Future (RFF).[3]

Table 2-1. Effects of Seven Potential Federal Renewable or Clean Electricity Standards

	25 Percent RES		25 Percent CES			20 Percent RES	
	With $50 ACP (EIA)	Without ACP (NREL)	With $50 ACP (RFF)	With $50 ACP (RFF)	With Natural Gas and $50 ACP(RFF)	With $25 ACP(RFF)	Without ACP (RFF)
Model Used	NEMS	ReEDS	NEMS	NEMS	NEMS	Haiku	Haiku
Final Year of Projection	2030	2030	2030	2030	2030	2035	2035
Unique Policy Features	Selected generation Receives triple credits; small utilities exempted	Selected generation receives triple credits; small utilities exempted	None	Credits for CCS and nuclear[a]	Credits for CCS and nuclear; partial credits for natural gas[b]	Energy baseline against which policy is measured includes hydropower	Energy baseline against which policy is measured includes hydropower
	Amount of Qualifying Generation in Final Year of Projection[c]						
Without Policy (Baseline)							
Billions of kilowatt hours	472	699	471	558	788	423	423
Percentage of electricity sales	10	14	10	12	18	9	9
With Policy							

	25 Percent RES			25 Percent CES		20 Percent RES	
	With $50 ACP (EIA)	Without ACP (NREL)	With $50 ACP (RFF)	With $50 ACP (RFF)	With Natural Gas and $50 ACP(RFF)	With $25 ACP(RFF)	Without ACP (RFF)
Billions of kilowatt hours	887	867	1,001	1,005	1,151	613	905
Percentage of electricity sales	20	18	22	22	26	14	20
	Policy-Induced Change in Generation in Final Year of Projection, by Source						
			(Billions of kilowatt hours)				
Renewable							
Wind	41	81	106	30	8	157	373
Biomass	360	14	410	125	117	33	100
Geothermal Energy	6	18	8	3	*	3	6
Solar Energy	8	55	7	6	4	0	0
Nonrenewable							
Coal with Carbon Capture and Storage	0	0	0	0	0	0	0
Coal	-257	-161	-298	-234	-228	-102	-275
Natural Gas	-150	1	-223	-216	-38	-84	-194
Nuclear Power	-31	0	-41	283	127	-63	-94
	Policy-Induced Change in Renewable Generation in Final Year of Projection, by Region						
			(Billions of kilowatt hours)				
West	50	76	96	19	5	74	174
Plains	115	20	156	59	51	92	154
East Central	65	38	85	31	28	5	59
Southeast	146	30	153	45	45	24	83
Northeast	39	4	41	10	-1	*	12
	Policy-Induced Change in CO_2 Emissions from Electricity Sector in Final Year of Projection						
Millions of Metric Tons of CO_2	-307	-150	-375	-307	-241	-138	-355

Table 2-1. (Continued)

	25 Percent RES			25 Percent CES		20 Percent RES	
	With $50 ACP (EIA)	Without ACP (NREL)	With $50 ACP (RFF)	With $50 ACP (RFF)	With Natural Gas and $50 ACP(RFF)	With $25 ACP(RFF)	Without ACP (RFF)
Percent	-12	-6	-14	-12	-9	-5	-14
	Average Policy-Induced Change in Electricity Prices						
	Between 2020 and 2030, by Region (Percent)						
West	1	**	1	**	**	**	-3
Plains	-3	-1	-4	**	1	1	2
East Central	2	-1	2	2	1	2	-1
Southeast	3	**	4	3	3	5	10
Northeast	3	**	2	1	2	-2	-4

Source: Congressional Budget Office based on analyses by the Energy Information Administration (EIA), the National Renewable Energy Laboratory (NREL), and Resources for the Future (RFF).

Notes: RES = renewable electricity standard; ACP = alternative compliance payment; CES = clean electricity standard; NEMS = National Energy Modeling System; ReEDS = Regional Energy Deployment System; CCS = carbon capture and storage; CO2 = carbon dioxide; * = between -0.5 billion and 0.5 billion kilowatt hours; ** = between -0.5 percent and 0.5 percent. The amounts shown here for alternative compliance payments are for the first year of the policy; those amounts rise over time with inflation.

[a] Facilities that began operations after the policy went into effect and produce electricity using either nuclear power or fossil fuel with CCS would receive one credit for each kilowatt hour of electricity generated; [b] Facilities that began operations after the policy went into effect and produce electricity using either nuclear power or fossil fuel with CCS would receive one credit for each kilowatt hour of electricity generated. Natural-gas-fired facilities without CCS that began operations after the policy went into effect would receive a fraction of a credit for each kilowatt hour of electricity generated; depending on the type of natural gas technology used, that fraction would vary from 0.33 to 0.59, reflecting each technology's rate of CO_2 emissions relative to that of a new pulverized-coal-fired boiler; [c] For RES policies, qualifying sources of electricity generation consist of renewable energy (except hydropower and municipal solid waste). For CES policies, qualifying sources of generation also include integrated gasification combined-cycle plants with carbon capture and storage, as well as advanced nuclear and (if noted) natural gas plants completed after the policy took effect. A few of the policies in this table include municipal solid waste as a qualifying source of generation; however, for consistency among the policies, CBO excluded that source from the numbers shown here.

CBO's aims in examining the seven analyses were to determine general trends in generation patterns resulting from RES and CES policies, to identify the effects that particular design features would have on policy outcomes, and to examine the extent to which models that reflect different assessments about the underlying costs of various technologies would predict different effects for similar policies. Those findings offer insights into the potential impact of any renewable or clean electricity standard that policymakers might consider.

Unless otherwise indicated, all of the policies examined here had the following features:

- Qualifying sources of energy included wind (both offshore and onshore), certain types of biomass, geothermal energy, and solar energy (both thermal and photovoltaic);
- Hydropower and electricity derived from municipal solid waste were not considered qualifying renewable sources and were excluded from the base to which the standard was applied—that is, utilities would not have to purchase credits for their sales of electricity that was generated by hydropower or municipal solid waste;
- Both existing and new sources of renewable generation were eligible for credits;
- Utilities could trade credits freely but could not bank or borrow them;
- The presence of states' renewable or clean electricity standards was accounted for in the baseline (that is, modelers increased the amount of renewable or clean electricity predicted to occur in the absence of a federal policy by the amount that would be mandated by state policies); and
- The federal policy would begin in 2012 and be phased in gradually. (All of the 25 percent RES and CES policies would be fully phased in by 2025; the two 20 percent RES policies would be fully phased in by 2020.)

Some of the policies included additional qualifying sources of electricity, such as ocean energy (power generated from tides or waves), landfill gas, and hydroelectricity from newly constructed facilities that met very specific restrictions. However, those sources were not projected to add significantly to renewable generation during the forecasting period—which ran through 2030 for analyses using the NEMS and ReEDS models and through 2035 for analyses using the Haiku model.

Each analysis reported regional changes in electricity generation and prices, but they defined regions in different ways. For example, ReEDS predicted changes at the state level, and NEMS produced estimates for 13 regions. For purposes of comparison, CBO adapted all of the models' results to the five regions used in the Haiku model (see Figure 2-1).[4]

Effects on Sources and Locations of U.S. Electricity Generation

The seven RES and CES policies examined here showed some similarities in their projected effects on the mix of energy sources used to generate electricity. Nevertheless, the effects varied significantly depending on the specific features included in the policies and the assumptions used in the models. Policy features such as an alternative compliance payment can cause otherwise identical policies to have widely varying outcomes. In addition, the three models used to analyze the policies incorporated different estimates of the cost of producing electricity from various sources in the coming decades. As a result of those differences, the three models would predict different outcomes even if they were used to analyze the exact same policy.

Similar Patterns in Findings

Most of the analyses concluded that the bulk of the increase in renewable generation caused by an RES or CES would come from wind and biomass (see Figure 2-2). Analyses conducted using NEMS predicted that, on average, roughly 80 percent of policy-induced increases in renewable generation would stem from increases in biomass generation and that most of the rest would come from wind generation. Analyses using Haiku, which incorporated different estimates of the availability of renewable resources, also found that most new generation would be in the form of biomass and wind, but in a different proportion than NEMS; they estimated that wind would account for nearly 80 percent of policy-induced increases in renewable generation and that biomass would account for the rest. The NREL analysis, conducted using ReEDS, estimated a significantly larger increase in geothermal or solar generation than the other analyses did. As discussed below, that outcome stems from specific assumptions in the ReEDS model about the relative cost of various forms of renewable generation, as well as specific provisions in the policy that ReEDS was used to analyze.

In general, the three models projected that most of the increase in renewable generation would occur in the Southeast, the Plains, and the West (see Figure 2-3 on page 21). Analyses using NEMS, which predicted that the bulk of new renewable generation would come from biomass, concluded that the largest increases would occur in the Plains and the Southeast. Analyses using Haiku, which predicted that most renewable generation would come from wind, found the largest increases in renewable generation occurring in the West and the Great Plains— areas that tend to have the best conditions for wind generation (see Figure 1-4 on page 11). The analysis conducted with ReEDS found new renewable generation more evenly distributed across the country. However, all of the models predicted that the Northeast would see the smallest increase in renewable generation.

The two analyses of a clean electricity standard predicted, not surprisingly, that because the CES would be met in part by increases in nuclear generation, a 25 percent CES would result in much less expansion of renewable generation than a 25 percent RES would (see Figure 2-2). For example, according to analyses by Resources for the Future, generation from wind and biomass would be 106 billion and 410 billion kilowatt hours higher, respectively, in 2030 as a result of a 25 percent RES but would be just 30 billion and 125 billion kilowatt hours higher, respectively, under a 25 percent CES without natural gas (see Table 2-1 on page 16). Allowing natural gas to qualify as a clean source under a CES would reduce the net increase in wind and biomass generation even more.

Effects of Policy Design Features on the Amount of Renewable or Clean Generation

Variations in the details of otherwise similar policies can create large difference in the policies' effects. The actual percentage of electricity generated from renewable or other clean sources under an RES or CES would most likely differ from the stated standard. The size of that difference would depend on specific design features—such as exemptions for certain utilities, preferences for selected technologies, and alternative compliance payments. In five of the seven analyses that CBO examined, the projected share of generation from qualifying sources was 3 to 7 percentage points less than the stated standard. In the other two analyses, the percentage of qualifying generation either met or exceeded the standard (for reasons explained below).

Exempting certain utilities from the standard, and excluding some types of generation from the base to which the standard was applied, would tend to

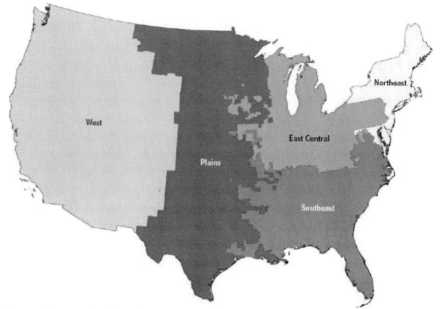

Sources: Congressional Budget Office; Resources for the Future.

Note: The five regions shown here correspond to the regions used in reporting results from the Haiku model used by Resources for the Future. The regions are aggregations of electricity markets, which do not necessarily correspond to state boundaries.

Figure 2-1. The Five Regions Used for Reporting Changes in Prices and Generation in This Study.

cause the percentage of generation provided by qualifying sources to be less than the standard.

Two of the versions of an RES examined here would exempt small generators, thereby reducing the share of electricity sales covered by the standard. Both EIA and RFF analyzed a 25 percent RES using NEMS; however, RFF found that the policy would induce more renewable generation than did EIA (see Figure 2-4), in part because EIA assumed that small utilities would be exempt from complying with the standard. Further, five of the seven policies would exclude existing generation from hydropower and municipal solid waste (which together account for 6.3 percent of current generation) from the base to which the standard was applied. That exclusion alone would mean that the share of generation provided by qualifying sources could not exceed 93.7 percent (100 − 6.3 percent) of the stated standard.

(Billions of kilowatt hours)

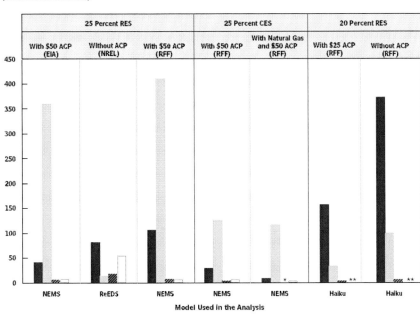

Source: Congressional Budget Office based on analyses by the Energy Information Administration (EIA), the National Renewable Energy Laboratory (NREL), and Resources for the Future (RFF).

Notes: RES = renewable electricity standard; ACP = alternative compliance payment; CES = clean electricity standard; NEMS = National Energy Modeling System; ReEDS = Regional Energy Deployment System; * = no significant change in generation using geothermal energy; ** = no change in generation using solar energy. The final year of the projection is 2030 for the 25 percent RES and the 25 percent CES and 2035 for the 20 percent RES.

Figure 2-2. Policy-Induced Change in Renewable Generation for the Final Year of the Projection, by Source.

RFF's analysis of a 20 percent RES was the only one that did not exempt sources from the base to which the policy was applied and, as a result, was the only policy in which the share of qualifying generation was equal to the standard, but only if an ACP was not included (see Figure 2-4).

Giving preference to certain technologies by awarding them more than one credit for each MWh of electricity that they produced would allow the standard to be met with a smaller rise in qualifying generation, increasing the extent to which the actual percentage of generation from qualifying sources would be less than the standard. The 25 percent RES policies modeled by the Energy Information Administration and the National Renewable Energy Laboratory (using the NEMS and ReEDS models, respectively) both contained a provision that would provide three credits for each MWh of electricity produced using solar photovoltaic technology. Because the ReEDS model included relatively optimistic assumptions about the cost of generating solar photovoltaic electricity (as discussed in the next section), that triple credit led to a much larger increase in solar generation in NREL's analysis than in EIA's analysis (see Figure 2-3 on page 21).

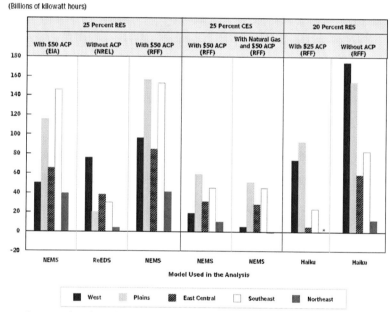

Source: Congressional Budget Office based on analyses by the Energy Information Administration (EIA), the National Renewable Energy Laboratory (NREL), and Resources for the Future (RFF).

Notes: RES = renewable electricity standard; ACP = alternative compliance payment; CES = clean electricity standard; NEMS = National Energy Modeling System; ReEDS = Regional Energy Deployment System; * = no significant change in generation in the Northeast. The final year of the projection is 2030 for the 25 percent RES and the 25 percent CES and 2035 for the 20 percent RES.

Figure 2-3. Policy-Induced Change in Renewable Generation for the Final Year of the Projection, by Region.

(Percent)

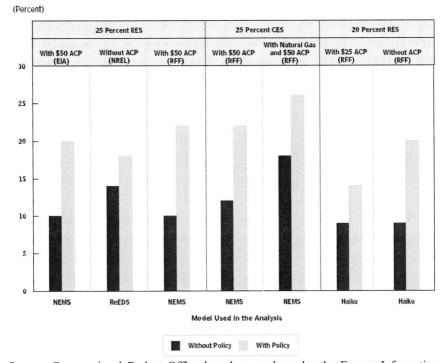

Source: Congressional Budget Office based on analyses by the Energy Information Administration (EIA), the National Renewable Energy Laboratory (NREL), and Resources for the Future (RFF).

Notes: RES = renewable electricity standard; ACP = alternative compliance payment; CES = clean electricity standard; NEMS = National Energy Modeling System; ReEDS = Regional Energy Deployment System.

The final year of the projection is 2030 for the 25 percent RES and the 25 percent CES and 2035 for the 20 percent RES.

Figure 2-4. Percentage of Qualifying Generation for the Final Year of the Projection, Without and With the Policy.

That increase in turn contributed to the substantial difference between the standard (25 percent) and the actual share of generation from renewable sources (18 percent) in NREL's analysis (see Figure 2-4). Providing additional credits for particular technologies would also alter the mix and location of qualifying electricity generation used to meet the standard.

Another policy feature that could cause the share of generation provided by qualifying sources to be much lower than the stated standard is allowing utilities to comply through an alternative compliance payment. Resources for the Future's analysis of a 20 percent RES (conducted using the Haiku model) demonstrates the potential effect of an ACP. The same 20 percent standard resulted in an actual share for renewable electricity of 14 percent when an ACP of $25 per MWh was included, compared with a 20 percent share when no ACP was included (see Figure 2-4). The lower that policymakers set an ACP relative to the anticipated price of a credit, the larger would be the difference between the actual percentage of qualifying generation and the standard, because utilities would have a greater incentive to pay the ACP instead of submitting credits. Most of the other analyses that CBO examined included ACPs, but those payments had little impact because they were set at a high enough level ($50 per MWh) that they affected compliance in few, if any, years of the projections.

Unlike the foregoing policy features, providing partial credits to certain technologies could cause the share of generation from qualifying sources to exceed the standard. For example, in Resources for the Future's analysis of a 25 percent CES in which electricity generated from natural gas would receive a fraction of a credit, the share of generation provided by all qualifying sources was projected to be slightly higher than the 25 percent standard once the policy was fully in place.[5] Providing partial credits for a particular technology would also create an upper limit on the use of that technology. For example, if an 80 percent CES program provided half a credit for each MWh of electricity generated from natural gas and a full credit for electricity from all other desig designated clean sources, the share of total generation provided by natural gas could not exceed 40 percent even if all other generation was from zero-emission sources.

Effects of Models' Differing Estimates of Costs

The future cost of generating electricity—whether from renewable or other clean sources or from conventional sources—is uncertain. Models that use different assumptions about costs will yield different predictions of the increase in renewable or clean generation that would be required to comply with a policy, as well as the mix of renewable or other clean sources that would result from a given standard. Differences between models' outcomes are an indication of the uncertainty about the actual effects of a policy.

The extent to which any particular RES or CES would increase renewable generation depends on the baseline against which the policy is measured—that is, on the amount of renewable generation that is predicted to occur in the future in the absence of the policy. Analyses in which the estimated cost of renewable generation is low relative to the cost of nonrenewable generation envision a larger share of renewable generation in the baseline. Thus, they predict that a smaller increase in renewable generation would be necessary to comply with a given standard. For example, the ReEDS model includes relatively low estimates of the cost of generating electricity from renewable sources, and it predicts that, in the absence of a standard, renewable generation will account for 14 percent of electricity sales in 2030—4 percentage points higher than in any of the other RES analyses examined. As a result, ReEDS would predict a smaller increase in renewable generation from any particular standard than other models would (see Figure 2-4).

In addition to the total change in renewable or clean generation, the mix of energy sources that each model predicts depends on many different factors. An increase in renewable or clean generation would tend to displace the most costly nonqualifying generation for which it could serve as a substitute. If all else was equal, such substitutions would be cheapest to make in places that required new capacity, because generators could add renewable or other clean capacity, rather than nonqualifying capacity, to meet that need but would not have to scrap existing generating capacity. Thus, differences in models' outcomes will depend on differences in assumptions about the costs of all types of generation, as well as on projections about the demand for new capacity in various regions. As a result, estimates of how the mix of generation would change because of a policy are uncertain, and estimates of the regional effects of a policy are even more uncertain.

Effects on Greenhouse Gas Emissions

An RES or CES would be likely to reduce greenhouse gas emissions, but the size of the reduction associated with a given standard is difficult to predict. It would depend in part on how much additional renewable or clean generation resulted from the standard. For example, although EIA's and NREL's analyses of a 25 percent RES showed similar shares of electricity being generated from renewable sources (20 percent and 18 percent, respectively), EIA predicted that the resulting decrease in CO_2 emissions in 2030 would be more than twice as large as the decrease projected by NREL (see Table 2-1 on page 16). That disparity stems mainly from differences in the baseline amount of renewable generation predicted by EIA's NEMS model and NREL's ReEDS model. The ReEDS model predicted 40 percent more renewable generation in the absence of a standard than the NEMS model did. Consequently, the analysis using ReEDS projected that the RES would lead to a smaller increase in renewable generation and hence a smaller reduction in CO_2 emissions.

The actual decrease in CO_2 emissions that would result from a policy would also depend on the type of generation being displaced by new renewable or clean generation. The reduction in emissions would be greater if renewable generation displaced coal-fired generation rather than natural-gas-fired generation, because generating a MWh of electricity from coal produces roughly twice as much CO_2 as generating the same amount of electricity from natural gas.[6] In general, biomass generation is somewhat more likely to displace coal-fired generation, and wind generation is somewhat more likely to displace natural-gas-fired generation (because of their locations and differences in their ability to provide base-load generation). Thus, if everything else is equal, models that predict that generators will rely on increases in biomass generation to comply with a policy will project larger reductions in CO_2 than will analyses that predict greater reliance on wind generation.[7]

Policymakers could set a more ambitious standard for the amount of generation to come from clean sources than for the amount to come from renewable sources. The reason is that a CES would allow for the use of technologies—such as nuclear power and coal- or natural-gas-fired generation with carbon capture and storage—that could probably be deployed on a large scale (setting aside concerns about health and environmental effects and costs, as well as technological uncertainty about carbon capture and storage) and that could provide a reliable source of base-load generation. As noted in Chapter 1, the potential for large-scale deployment of renewable technologies is much more limited, because such technologies are typically well suited to only certain parts of the country and, in the case of wind and solar generation, are dependent on fluctuating conditions.

Effects on Prices for Electricity

Either an RES or a CES would raise the overall cost of producing electricity in the United States. Without such a standard, generators in competitive electricity markets would choose the mix of sources that maximized their profits, and in regulated markets, regulators would tend to require a mix that minimized the cost of electricity production. An RES or CES would induce them to alter that mix and produce electricity in a more costly manner (not accounting for any environmental or other benefits of changing the mix).

Factors Affecting Electricity Prices in Different Areas

The change in the generation mix prompted by an RES or CES, and the fact that utilities would have to buy (or in some cases could sell) credits, would alter electricity prices throughout the country. Regional price changes would depend on changes in regional production costs, the amount of electricity and credit revenue flowing into and out of regions, and the manner in which prices were determined in each location.

The majority of states have regulated electricity markets (known as "cost-of-service" markets). In those areas, policy-induced changes in electricity prices would reflect changes in the average cost of producing electricity in the region as well as purchases and sales of credits. Prices would go up in regulated areas if average production costs increased because of generators' efforts to use qualifying sources to meet the region's compliance obligations or if the region became a net purchaser of credits. Prices could fall in a regulated region if it became a net seller of credits. For example, electricity prices could fall in a regulated market under a 25 percent RES if, before the policy, 30 percent of the region's electricity came from wind. In that case, existing wind generators would be able to sell more credits than the region would need to comply with the policy, and the revenue from the additional sales would be passed on to electricity customers in the form of lower prices.

Electricity prices could also rise or fall in the 14 states that have restructured their markets to give consumers access to competitive interstate electricity markets. Electricity prices in those competitive markets reflect the cost of the "marginal" supply of electricity (that is, the incremental cost of the most expensive generation used to meet demand). In those markets, the effects of an RES or CES on retail electricity prices would depend on how the increased reliance on renewable sources affected the incremental cost of the marginal supplier. Average costs would be higher with a standard, but those averages could include high fixed costs and low variable (and thus, low incremental or marginal) costs for some generation sources. Thus, prices based on marginal costs could either rise or fall in competitive electricity markets under an RES or CES.

If a standard caused electricity prices to rise in a region, the costs of the policy would be borne by local consumers of electricity; but if prices fell in a region, the costs would be borne by electricity producers in that region or by consumers in *other* regions (areas that were net buyers of credits would subsidize production in areas that were net sellers of credits). Costs would tend to be borne by producers in places where wholesale electricity prices were set in competitive markets and where the policy altered the generation mix in such a way that the marginal cost of producing electricity in the region declined. That outcome was predicted for the Northeast under both of the 20 percent RESs analyzed using the Haiku model: Additional wind generation was predicted to displace natural-gas-fired generation there, causing the marginal cost of producing electricity in the region to fall.

Although producers could bear the costs of a reduction in prices in a region for some time (until existing plants were retired), such an outcome could not be sustained over the very long run. Policy-induced decreases in electricity prices in a region would cause the value of existing capital for nonqualifying generation to decline, because the present value of the anticipated revenue from using that capital stock would be lower. That decline in value would cause less new capital to enter the industry and could cause existing capital to be retired sooner than it would be otherwise. Having less generating capacity, in turn, would decrease the supply of electricity and eventually lead to higher electricity prices. Thus, consumers would bear the costs of meeting the standard in the very long run.

Results of Specific Analyses

All of the analyses that CBO examined reported policy-induced changes in retail electricity prices; however, they reported the changes for different years and different geographic areas. For purposes of comparison, CBO looked at average changes in retail electricity prices between 2020 and 2030—the period when most of the policies would have the largest impact on prices—and mapped the price results into the five regions used above to compare the estimated effects of policies on the sources and locations of electricity generation (see Figure 2-1 on page 19).

The RES and CES policies examined in this study tended to produce small increases in average retail electricity prices in the United States. Those price effects varied significantly, however, among models and policies. For example, the two 25 percent RES policies modeled by EIA and RFF and the 25 percent CES that excluded natural gas were predicted to reduce electricity prices in the Plains region but to increase prices in the other four regions (see

(Percentage change)

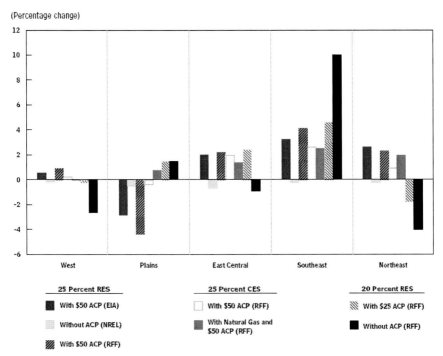

Source: Congressional Budget Office based on analyses by the Energy Information
   Administration (EIA), the National Renewable Energy Laboratory (NREL), and
   Resources for the Future (RFF).
Notes: RES = renewable electricity standard; ACP = alternative compliance payment;
   CES = clean electricity standard.
The regions used in this figure correspond to those shown in Figure 2-1 on page 19.
   For more information about the seven electricity standards analyzed, see Table 2-1
   on page 16.

Figure 2-5. Average Change in Electricity Prices Between 2020 and 2030, by Region,
Under Seven Potential Federal Electricity Standards.

Figure 2-5). Including natural gas as a qualifying source in the CES caused
electricity prices to be virtually unchanged in the West but led to higher prices
in the four remaining regions, including the Plains.

   The two 20 percent RES policies (with and without an ACP) modeled
using Haiku produced their largest price increases in the Southeast and their
largest price decreases in the Northeast. Including the ACP significantly
dampened the price changes that would otherwise result from that policy.

The 25 percent RES modeled by NREL using ReEDS produced very small decreases in electricity prices in all five regions. As described above, the ReEDS model projected more renewable generation in the baseline than other models did and thus required less change in the generation mix—and less additional cost—to meet any particular standard. That lower additional cost, however, does not explain ReEDS's projection of price declines in all regions. Prices would drop everywhere only if producers, rather than consumers, bore the cost of the policy. But as noted above, such losses would manifest themselves as a reduction in the value of existing capital in the electricity-generating sector and would ultimately decrease the amount of generating capacity. That decrease, in turn, would eventually cause prices to rise, meaning that the cost of the policy would ultimately be borne by electricity consumers.

The large variation in price changes found by different models and for different policies indicates the substantial uncertainty about regional price effects. The price changes predicted for a given policy in a given area are highly dependent on the baseline generation mix projected for that area, the investment in qualifying sources that takes place, and the manner in which prices are determined. Thus, although the price effects of any given RES or CES will inevitably vary among regions, the pattern of that variation is difficult to predict with certainty.

Interaction of a Federal Standard with Existing State Standards

The wide variety of renewable and clean electricity standards that exist at the state level make it harder to assess the impact of imposing a federal RES or CES. Moreover, enacting a federal standard would alter the effects of states' policies. The combined effects of federal and state standards would depend on details of all of the standards, such as whether a single MWh of renewable or other clean electricity generation could qualify for both a federal and a state credit. Because of the additional constraints that state standards could impose on the mix of qualifying generation, any desired increase in the overall percentage of electricity produced from renewable or other clean sources in the United States could be achieved at a lower cost through a well-designed federal standard than through a combination of a federal standard and numerous state standards.

Effects on the Total Amount of Qualifying Generation

If a federal RES or CES existed alongside current state standards, a central question in predicting the impact on the total amount of renewable or clean electricity generation is whether the policies would overlap (so the same unit of power generation could be used to comply with both standards) or whether they would be cumulative (so more qualifying generation would be needed to satisfy both policies).

If the same MWh of qualifying electricity generation could be used to comply with both a federal and a state standard, the total amount of renewable or clean generation in the United States would be determined by either the combined effect of the state standards or by the federal standard—whichever was more stringent. That situation would arise if state governments and the federal government operated their own programs, issuing their own credits and monitoring compliance with their particular standard. That approach would allow both levels of government to define their own program rules and to ensure their own standards for monitoring and enforcement; however, it would entail duplicative administrative and enforcement efforts.

If, by contrast, a MWh of qualifying generation could not be used to comply with both the federal standard and a state standard, total renewable or clean generation nationwide would be greater than either the amount required by the federal standard or the aggregate amount mandated by state standards. That situation would arise if state-issued credits could be used only to comply with state programs and federal-issued credits only with the federal program, and if each qualifying MWh of electricity could be issued a credit under either program but not both. For example, if utilities in Nevada simultaneously had to comply with that state's 15 percent RES and a 5 percent federal RES in 2015, and if they could not use the same unit of renewable generation to meet both standards, they would have to obtain enough credits to cover 20 percent of their electricity sales, a percentage greater than either the state or the federal RES.

The existence of numerous state standards makes it more difficult to predict not only the total amount of qualifying generation that would occur in the absence of a federal policy but also the incremental increase in such generation that a federal RES or CES would bring about. As discussed above, the renewable or clean generation induced by state standards is typically built into baseline projections of the mix of generation that would occur in the absence of a federal RES or CES. That baseline is thus heavily dependent on state policies that may change over time. More-stringent state standards diminish the incremental impact of any particular federal standard.

Effects on Credit Prices

State standards that motivate more renewable or clean generation in a state than would have occurred in the absence of those standards would tend to lower the price of a federal credit and the cost of complying with the federal program, provided that a single MWh of generation could qualify for credits under both programs. In essence, each MWh of renewable or other clean electricity that is generated in a state as a result of the state's standard simultaneously would produce a credit for use in the federal program at no additional cost. That "free" credit would lower the cost of meeting the federal standard and the price of credits traded in the federal program—but only because the cost of producing that credit would be attributed to the cost of meeting the state program and be reflected in the price of state credits.

Just as the existence of state programs would affect the price of credits traded in a federal program, enactment of a federal program would affect the price of credits traded in state programs. For example, if a federal standard induced more qualifying generation in a particular state than required under the state standard, the federal program would drive down the price of that state's credits to zero.

Effects on the Overall Cost of Increasing Renewable or Clean Generation

Regardless of what increase policymakers wanted to achieve in the share of U.S. electricity generated from renewable or other clean sources, it would cost less to produce that increase with an appropriately designed federal standard than with a combination of federal and state standards. By itself, a federal standard would create a national credit-trading program that would allow market forces to determine where and how investments in qualifying sources of electricity generation would occur. The existence of state standards, however, would constrain that mix, potentially forcing higher-cost generation to be used for compliance than would otherwise be the case.

3. CONSIDERATIONS IN DESIGNING A COST-EFFECTIVE RENEWABLE OR CLEAN ELECTRICITY STANDARD

The overall cost of meeting a renewable or clean electricity standard cannot be known with certainty, but policymakers could help minimize those costs by adding various design features, such as allowing credits to be traded freely among interested parties, letting as many energy sources as possible generate credits, gradually phasing in the standard, and permitting utilities to shift credits between years through banking and borrowing. The seven analyses that the Congressional Budget Office examined in Chapter 2 assumed that credits could be traded and that a wide variety of sources, but not all potential sources, would be allowed to generate credits. (For example, none of the analyses assumed that electricity generated by existing natural-gas-fired plants would be eligible for partial credits.) Finally, all of the policies analyzed assumed that the standards would be phased in gradually but that firms would not be allowed to bank or borrow credits.

If the objective is to reduce carbon dioxide emissions, a program in which the government set a nationwide cap on such emissions and created tradable emissions allowances could offer a lower-cost method of achieving that goal than either an RES or a CES (see Box 3-1). Such a cap-and-trade program could be more comprehensive than an electricity standard. It would also provide a direct incentive to curb CO_2 emissions in many different ways, could motivate low-cost reductions throughout the entire economy, and could guarantee that emissions stayed below a defined level. By contrast, an RES or CES would provide direct encouragement for the use of renewable or other clean energy sources, but CO_2 emissions would be reduced only as a by-product of utilities' compliance with the standard.

Effects of Credit Trading on Costs

Because opportunities for expanding renewable and clean electricity generation vary greatly among regions, credit trading would enable the nation to meet an RES or CES by using the lowest-cost opportunities. (That variation reflects, among other things, differences in the availability of renewable resources and the need for additional generating capacity.)

As a further step, allowing entities other than generators and utilities— such as financial firms—to trade credits could provide substantial cost savings.

Box 3-1. The Cap-and-Trade Alternative
for Decreasing CO_2 Emissions

Neither a renewable electricity standard (RES) nor a clean electricity standard (CES) could achieve a desired reduction in carbon dioxide (CO_2) emissions at as low a cost as a "cap-and-trade" program could. Under such a program, the federal government would set increasingly tight annual limits on greenhouse gas emissions (including CO_2, the primary greenhouse gas emitted as a result of human activity) over the course of several decades. The government would distribute rights (or allowances) for those emissions by either selling them, possibly in an auction, or by giving them away. Companies that would be subject to the cap—such as electricity generators, oil importers, and natural gas producers—would be permitted to buy and sell allowances after the initial distribution. They could also shift allowances over time to some degree by "banking" unused allowances for future use or by "borrowing" allowances designated to permit emissions in future years. The price of allowances would rise to the level necessary to ensure that the limit on cumulative emissions over the life of the policy (implied by the annual caps) was met.

A federal cap-and-trade program has been operating in the United States since 1995 to limit emissions that cause acid rain, and the European Union and a number of U.S. states have implemented cap-andtrade programs to curb greenhouse gas emissions.[1] The 111th Congress debated, but did not pass, a federal cap-and-trade program for greenhouse gas emissions.[2]

A cap-and-trade program could cover the whole economy, or it could be limited to the electricity sector. The wider the coverage, the more cost-effective the policy would be. Even a cap-and-trade program that was limited to the electricity sector could achieve a given reduction in CO_2 emissions at a lower cost than an RES or CES could. A cap-and-trade program that covered the entire economy could achieve that cut at an even smaller cost.

The main advantage of a cap-and-trade program over an electricity standard is that it could provide a direct and comprehensive incentive to decrease the CO_2 emissions caused by burning fossil fuels. An economywide cap-and-trade program would cause the price of all goods and services to rise in proportion to the amount of emissions associated with their production and use. As a result, companies and households throughout the economy would have an incentive to take actions that cut emissions.

For example, electricity generators would have an incentive to make efficiency improvements to their boilers, switch to low-emitting fuels, and conserve energy in the operation of their facilities (such as by installing more-efficient lighting). Households, manufacturers, businesses, government agencies, and other entities would face a similar incentive: They would save money by altering their production and consumption in ways that reduced emissions, including by driving less or purchasing vehicles that have greater fuel efficiency. (Such activities would decrease the use of fossil fuels in the transportation sector, which emits about 40 percent of all CO_2 emitted in the United States.) Besides being direct and comprehensive, the incentive that entities faced to change their production processes or consumption patterns would be directly proportional to the cut in emissions that those changes would bring about. Finally, a cap-andtrade program could ensure that emissions remained below a desired level.

In contrast, the incentive to reduce CO_2 emissions that would result from an RES or CES would be neither comprehensive nor direct. An electricity standard would provide a direct economic incentive to increase generation from renewable or other clean energy sources; CO_2 emissions would be reduced only as a by-product of achieving that goal, and the scale of the reduction would be uncertain.

Moreover, the design of an RES or CES would determine the extent to which the size of the incentive to produce electricity from qualifying sources (rather than nonqualifying ones) was linked to the size of the cut in emissions that such a change would bring about. A policy in which a wide variety of sources qualified for credits and in which the amount of credits was related to a source's CO_2 emissions would more closely align financial incentives with actual cuts in emissions than would a policy lacking those design characteristics. For example, if a CES included partial credits for both existing and new natural-gas-fired generation, generators would receive a larger financial reward if they substituted a megawatt hour of emission-free generation for a megawatt hour of generation from a high-emitting source, such as coal, than from a low-emitting source, such as natural gas. The former substitution would result in nearly twice the emission reductions as the latter.

Even in a comprehensive CES, however, financial incentives and emission cuts could be aligned only imperfectly. Although regulators could assign credits on the basis of a given technology's average emissions, actual performance (and thus emissions) would vary around those averages.

Further, regulators would need to update the amount of credits awarded for various technologies as the technologies changed over time.

Having a broader range of participants would increase the liquidity of the market for credits, making it easier for utilities and generators to identify other parties with whom to trade and to complete large transactions without affecting the market price of credits. Another benefit of broad participation would be to improve the quality of the information underlying credit prices. Like traders in any market, active credit traders would have a financial incentive to study the market and the industries underpinning it, because their profits from providing liquidity would depend on their ability to forecast the future price of credits. As a result, those traders would bring to the bargaining table up-to-date information about the factors that would determine credit prices, such as the demand for electricity and the cost of adding qualifying generation.

Allowing a wide variety of entities to trade credits would be likely to spawn a market for credit derivatives—financial contracts whose value would depend on the future price of credits. Utilities, electricity generators, and other participants in the credit market could use derivatives to protect themselves from changes in prices (just as farmers use derivatives on agricultural commodities to lock in prices for their crops before the crops are harvested or even planted).

The availability of derivatives would lower a utility's compliance costs by giving it a relatively low cost and convenient way to reduce uncertainty about the cost of acquiring credits in the future. In the absence of derivatives, utilities' main option for avoiding that uncertainty would be to buy and hold large amounts of credits themselves.

At the same time, observers worry that permitting third parties to trade credits and allowing the use of derivatives could foster complex or opaque transactions, price bubbles, market manipulation, or other instabilities that could raise the cost of complying with the standard and harm the broader economy. In the case of a proposed national cap-and-trade program, that potential spurred calls to prohibit or otherwise limit certain types of market participants or transactions. An earlier CBO analysis of such proposals concluded that less restrictive limits would generally have a greater chance of addressing observers' concerns, with fewer negative effects, than outright prohibitions would.[1]

Effects of Maximizing Compliance Options on Costs

A CES could bring about any specified reduction in CO_2 emissions at a lower cost than an RES would: By allowing a wider set of qualifying sources of electricity, it would provide more options for reducing emissions. If technologies that emit significant amounts of CO_2 were included as qualifying sources, the credits that each type of generation earned per megawatt hour could be adjusted to account for the differences in emissions. Such weighting would help induce cost-effective cuts in CO_2 emissions because it would provide the largest financial incentive for generators to rely on sources with the lowest emissions. For example, if the price of a credit was $20, each MWh of electricity produced at new nuclear plants, which do not emit any CO_2, could receive a full credit, worth $20, whereas each MWh of electricity produced at a new natural-gas-fired plant, which emits about half as much CO_2 as electricity generated by coal, could receive a half credit, worth $10. As a result, all else being equal, generators would be more likely to build a new nuclear plant (earning $20 in credit revenue for each MWh of electricity it produced) than a new natural gas plant (earning only $10 in credit revenue for each MWh of electricity it produced).

Another way to increase compliance options would be to award credits for qualified electricity savings (cuts in electricity use that result from efficiency improvements that would not have occurred in the absence of the policy). Some of the electricity standards proposed in the 111th Congress would have let utilities, generators, and other entities obtain credits for qualified improvements in energy efficiency.

In theory, allowing efficiency improvements to earn credits could decrease the cost of both complying with a CES and reducing CO_2 emissions. Because such improvements reduce the total amount of electricity generated, they would lessen the increase in qualifying generation that would be necessary to meet the standard. (In other words, electricity savings would shrink the base to which the standard was applied.) Further, some energy-efficiency improvements would represent low-cost ways of cutting CO_2 emissions.

In practice, however, measuring the amount of energy saved through an efficiency improvement and ascertaining that the savings would not have happened in the absence of the policy could pose administrative challenges.[2] Calculating how much energy was saved because of an efficiency improvement would require determining a baseline measure of the energy consumption that would have occurred without the efficiency upgrade. Under some Congressional proposals, qualifying energy-efficiency credits would be determined by comparing a facility's energy use before and after an efficiency improvement (for example, comparing the electricity consumption at a facility that installed more-efficient lighting with its average electricity consumption in the five years before the improvement) or, in the case of a generating plant, by comparing the amount of fossil fuel used to produce a kilowatt hour of electricity before and after a facility upgrade.[3]

Even if such energy savings could be measured accurately, it would be hard to gauge whether generators and utilities would have made such improvements in the absence of the program—that is, whether the energy savings were truly "additional" or whether entities were receiving credits for improvements they would have found it profitable to make anyway. By allowing utilities to submit energy-efficiency credits in lieu of credits earned by generating clean or renewable electricity, policymakers essentially would be agreeing to a relaxation of the CES or RES in return for a reduction in either the consumption of electricity (as in the case of the replacement lighting) or the emissions per MWh of electricity generated from a given source (as in the case of the facility upgrade). But if the activities generating the energy-efficiency credits would have been undertaken even in the absence of the policy, awarding credits for them would result in a relaxation of the standard but not yield a return. (In contrast, in the absence of energy-efficiency credits, the standard would be met regardless of whether generators would have found it profitable to install wind or biomass generation without a standard.) For all of those reasons, allowing energy efficiency to earn credits might undermine the credibility of the standard and could increase the net cost of implementing an RES or CES despite the two types of cost savings noted above.

Effects of Timing on Costs

An RES or CES would be less expensive to meet if it was phased in gradually and then held constant at the desired level for a long period. Gradually phasing in the standard would increase the extent to which generators could comply through substituting new qualifying generating capacity for their planned replacements or expansions of nonqualifying capacity. In contrast, compliance costs would be higher if the standard was phased in so quickly that generators did not have time to recoup their invest-ments in existing capital. Signaling credibly that the policy would remain in place over a period of several decades would make it easier to finance a renewable or clean generating facility by providing a degree of certainty about how long the facility would receive credits. Again, in contrast, if the policy was not expected to persist well into the future, the cost of adding qualifying capacity would be high because generators would expect that they could obtain credits for only a few years. Providing certainty about the policy for an extended period would also let utilities enter into long-term contracts with generators for purchases of electricity and credits.[4]

Allowing utilities to bank and borrow credits would affect the rate at which the amount of qualifying generation grew by letting utilities tailor the timing of their compliance to the actual cost of increasing qualifying generation. Banking would lower costs by allowing companies to save excess credits obtained in years when renewable or clean generation exceeded the amount needed to meet the standard and then use the credits in later years. Borrowing credits expected to be earned in the future could be helpful in lowering compliance costs if it prevented the need for existing generating facilities to be retired prematurely. The standard would not ultimately be met, however, if firms that borrowed credits declared bankruptcy and the borrowed credits were never repaid. Policymakers could reduce that possibility by restricting the number of credits that utilities could borrow—to 10 percent of their compliance obligation, for example— and by limiting how far in advance credits could be borrowed.

In essence, phasing in a standard would let policymakers balance the desire to change the generation mix quickly with the fact that faster changes entail higher costs. The cost side of that trade-off would be based on projections of future conditions. Giving firms the ability to bank and borrow credits would let them modify the time path selected by policymakers. Those modifications would be based on unfolding information about the cost of changing the generation mix at various points in time.

End Notes for Introduction

[1]. See Pew Center on Global Climate Change, "Renewable and Alternative Energy Portfolio Standards" (August 7, 2011), www.pewclimate.org/what_s_being_done/ in_the_states/rps. cfm.

[2]. For a more detailed discussion of that point, explained in the context of a cap-and-trade program for greenhouse gas emissions, see Congressional Budget Office, *Managing Allowance Prices in a Cap-and-Trade Program* (November 2010).

[3]. For a discussion of those implications in the context of a cap-and-trade program, see Congressional Budget Office, *The Economic Effects of Legislation to Reduce Greenhouse-Gas Emissions* (September 2009).

[4]. See R. Wiser, K. Porter, and R. Grace, "Evaluating Experience with Renewables Portfolio Standards in the United States" (Lawrence Berkeley National Laboratory, March 2004), http://eetd.lbl.gov/ea/ems/reports/54439.pdf.

[5]. Geothermal energy is considered to be sustainable because the heat extracted is small compared with the Earth's heat content.

[6]. Geothermal power provides the most extreme example of this constraint. Conventional methods may be used to generate electricity from geothermal energy at a cost that is competitive with coal-fired generation; however, the locations at which that can take place are so limited that geothermal power currently accounts for less than 1 percent of U.S. electricity generation and is not expected to grow significantly over the next several decades. Enhanced methods might allow for greater reliance on geothermal energy, but successful research and development would be necessary to make those enhanced methods cost-competitive.

[7]. Charles Ebinger, remarks given at the Brookings Institution seminar "Energy and Climate Change 2010: Back to the Future," Washington, D.C., May 18, 2010, www.brookings. edu/~/media/Files/events/2010/20100518_energy_climate/20100518_energy_climate_chan ge_II.pdf.

[8]. For a discussion of siting and cost-allocation issues, see Stan Mark Kaplan, *Electric Power Transmission: Background and Policy Issues*, CRS Report for Congress R40511 (Congressional Research Service, April 14, 2009); and Mason Willrich, "Electricity Transmission Policy for America: Enabling a Smart Grid, End-to-End" (seminar presentation to the Center for Environmental Public Policy, Goldman School of Public Policy, University of California, Berkeley, November 18, 2009), http://gspp.berkeley. edu/programs/docs/CEPP_Willrich_111809.pdf.

[9]. See Department of Energy, Energy Information Administration, "2016 Levelized Cost of New Generation Resources from the Annual Energy Outlook 2010" (no date), www.eia.doe.gov /oiaf/aeo/pdf/2016levelized_costs_aeo2010.pdf.

[10]. Failure to account for that fact can lead to misleading comparisons between the market value of wind or solar plants and conventional alternatives. See Paul L. Joskow, *Comparing the Costs of Intermittent and Dispatchable Electricity Generating Technologies* (Massachusetts Institute of Technology, Center for Energy and Environmental Policy Research, September 2010).

[11]. See Congressional Budget Office, *The Impact of Ethanol Use on Food Prices and Greenhouse-Gas Emissions* (April 2009).

[12]. See Kelsi Bracmort, *Biomass Feedstocks for Biopower: Background and Selected Issues*, CRS Report for Congress R41440 (Congressional Research Service, October 6, 2010), www.fas.org/sgp/crs/misc/R41440.pdf.

[13]. See Congressional Budget Office, *The Impact of Ethanol Use on Food Prices and Greenhouse-Gas Emissions*.

[14]. That phenomenon is so common that energy modelers often include a "technological optimism factor" to compensate for early underestimates of final costs. See Department of Energy, Energy Information Administration, "The National Energy Modeling System: An Overview—Electricity Market Module" (October 2009), www.eia.doe.gov/oiaf/aeo/ overview/electricity.html.

[15]. That approach would also mean that the reduction in CO_2 emissions induced by the policy would occur slowly.

[16]. See Department of Energy, Energy Information Administration, *Annual Energy Review 2009* (August 2010), Table 8.1, www.eia.gov/emeu/aer/pdf/pages/sec8_5.pdf; and Pew Center on Global Climate Change, "Electricity Overview" (June 2011), www.pewclimate.org/ technology/overview/electricity.

End Notes for Box 1.1.

[1]. See Department of Energy, Energy Information Administration, "Average Retail Price of Electricity to Ultimate Customers: Total by End-Use Sector" (revised April 2011), www.eia.gov/cneaf/electricity/epm/table5_3.html.

[2]. See Phillip Brown and Molly F. Sherlock, *ARRA Section 1603 Grants in Lieu of Tax Credits for Renewable Energy: Overview, Analysis, and Policy Options,* CRS Report for Congress R41635 (Congressional Research Service, February 8, 2011).

[3]. Ibid.

[4]. See Department of Energy, Energy Information Administration, *Annual Energy Outlook 2011, with Projections to 2035* (April 2011), www.eia.gov/forecasts/aeo/.

End Notes for "Potential Effects of a Renewable or Clean Electricity Standard"

[1]. Department of Energy, Energy Information Administration, *Annual Energy Outlook 2011, with Projections to 2035* (April 2011), www.eia.gov/forecasts/aeo/.

[2]. For descriptions of each analysis, see Department of Energy, Energy Information Administration, *Impacts of a 25-Percent Renewable Electricity Standard as Proposed in the American Clean Energy and Security Act Discussion Draft* (April 2009), www.eia.gov/oiaf/servicerpt/acesa/index.html; Patrick Sullivan and others, *Comparative Analysis of Three Proposed Federal Renewable Electricity Standards,* Technical Report NREL/TP-6A2-45877 (National Renewable Energy Laboratory, May 2009), www.nrel.gov/docs/fy09osti/45877.pdf; and Alan J. Krupnick and others, *Toward a New National Energy Policy: Assessing the Options* (National Energy Policy Institute and Resources for the Future, November 2010), http://nepinstitute.org/publications/toward-a-new-national-energy-policy-assessing-the-options/.

[3]. For documentation on NEMS, see Department of Energy, Energy Information Administration, "The National Energy Modeling System: An Overview" (October 2009), www.eia.gov/oiaf/aeo/overview/index.html; on ReEDS, see National Renewable Energy Laboratory, "Regional Energy Deployment System" (no date), www.nrel.gov/analysis/reeds/; and on Haiku, see Anthony Paul and Dallas Burtraw, *The RFF Haiku Electricity Market Model* (Resources for the Future, June 2002), www.rff.org/RFF/Documents/RFF-RPT-haiku.pdf.

[4]. Because the smaller geographic areas used in ReEDS and NEMS did not have boundaries that corresponded to those of the larger areas used in Haiku, CBO constructed weighted averages of the smaller areas such that their population matched that of the five regions in the Haiku model in the initial period of each projection.

[5]. That analysis also concluded that the use of natural gas would decline because of the policy even though utilities would comply in part by purchasing credits for electricity generated at new natural gas plants. That outcome illustrates the potential efficiency cost associated with providing credits only for generation from new, rather than all, natural gas plants. The policy would give utilities an incentive to invest in new natural gas plants but not an incentive to operate their existing plants more than they would otherwise. Thus, the policy could cause new plants (which would generate both electricity and credits) to be built while

existing plants were being used less. Providing credits for existing sources could solve that problem, but, as described in Chapter 1, it would also have significant distributional effects that policymakers might find undesirable.

[6]. See Alan J. Krupnick and others, *Toward a New National Energy Policy: Assessing the Options* (National Energy Policy Institute and Resources for the Future, November 2010), p. 99, http://nepinstitute.org/publications/toward-a-new-nationalenergy-policy-assessing-the-options/.

[7]. Some studies suggest that renewable electricity standards would be more likely to reduce use of natural gas than of coal. See Krupnick and others, *Toward a New National Energy Policy,* p. 100; and Carolyn Fischer and Louis Preonas, "Combining Policies for Renewable Energy: Is the Whole Less Than the Sum of the Parts?" *International Review of Environmental and Resource Economics,* vol. 4, no. 1 (June 2010), pp. 51–92.

End Notes for "Considerations in Designing a Cost-Effective Renewable or Clean Electricity Standard"

[1]. See Congressional Budget Office, *Evaluating Limits on Participation and Transactions in Markets for Emissions Allowances* (December 2010).

[2]. Similar challenges would arise in the determination of qualifying "offsets" under a cap-and-trade program; see Congressional Budget Office, *The Use of Offsets to Reduce Greenhouse Gases,* Issue Brief (August 2009).

[3]. Those approaches were included in the Renewable Electricity Promotion Act of 2010 (S. 3813) and the Practical Energy and Climate Plan Act of 2010 (S. 3464).

[4]. See Ryan Wiser and others, "The Experience with Renewable Portfolio Standards in the United States," *Electricity Journal,* vol. 20, no. 4 (May 2007), pp. 8–20.

End Notes for Box 3-1.

[1]. For more about the U.S. cap-and-trade program for emissions that cause acid rain, see Dallas Burtraw and others, "Economics of Pollution Trading for SO2 and NOx," *Annual Review of Environment and Resources,* vol. 30 (2005), pp. 253–289. For a discussion of the European cap-and-trade program, see A.D. Ellerman and P. Joskow, *The European Union's CO2 Cap-and-Trade System in Perspective* (Pew Center on Global Climate Change, 2008).

[2]. For more details of how a federal cap-and-trade program for CO2 emissions might work, see Congressional Budget Office, *The Economic Effects of Legislation to Reduce Greenhouse-Gas Emissions* (September 2009) and *Issues in the Design of a Capand-Trade Program for Carbon Emissions* (November 2003).

In: Renewable Electricity Generation   ISBN: 978-1-61942-678-8
Editors: L. G. Irwin and R. Macias   © 2012 Nova Science Publishers, Inc.

Chapter 3

ANALYSIS OF IMPACTS OF A CLEAN ENERGY STANDARD[*]

The United States Energy Information Administration

PREFACE

This report responds to a July 2011 request to the U.S. Energy Information Administration (EIA) from Chairman Ralph M. Hall of the U.S. House of Representatives Committee on Science, Space, and Technology, for an analysis of the impacts of a Clean Energy Standard (CES). The request, as outlined in the letter included in Appendix A, sets out specific assumptions and scenarios for the study.

INTRODUCTION

This report responds to a request from Chairman Ralph M. Hall for an analysis of the impacts of a Clean Energy Standard (CES). The request, as outlined in the letter included in Appendix A, sets out specific assumptions and scenarios for the study.

---

[*] This is an edited, reformatted and augmented version of The United States Energy Information Administration publication, as requested by Chairman Hall, dated October 2011.

Background

A CES is a policy that requires covered electricity retailers to supply a specified share of their electricity sales from qualifying clean energy resources. Under a CES, electric generators would be granted clean energy credits for every megawatt-hour (MWh) of electricity they produce using qualifying clean energy sources. Utilities that serve retail customers would use some combination of credits granted to their own generation or credits acquired from other generators to meet their CES obligations. Generators without retail customers or utilities that generated more clean energy credits than needed to meet their own obligations could sell CES credits to other companies.

The impact of a CES will be sensitive to its design details and to assumptions made regarding the cost of the different fuels and technologies that can be used for electricity generation. Chairman Hall's request asks for an evaluation of a particular CES under a variety of alternative assumptions regarding the costs of generation fuels and technologies.

The CES specified by Chairman Hall, hereinafter referred to as the Hall CES (HCES), has the following characteristics:

- Eligible resources to meet the HCES target include: hydroelectric, wind, solar, geothermal, biomass power, municipal solid waste, landfill gas, nuclear, coal-fired plants with carbon capture and sequestration, and natural gas-fired plants with either carbon capture and sequestration or utilizing combined cycle technology.
- Generators earn 0.5 MWh of compliance credits for every 1 MWh of generation from a combined cycle plant that burns natural gas, and 0.9 MWh of compliance credits for every 1 MWh of generation from coal- or gas-fired generation with carbon capture and sequestration. All other HCES-qualified resources earn one HCES credit for every MWh of generation.
- Generation using qualified resources from either new or existing plants in any economic sector can receive HCES credits.
- The HCES target starts from an initial share of 44.8 percent (qualified generation as a percent of sales) in 2013 and rises linearly to 80 percent in 2035. Beyond 2035, the target remains at 80 percent.
- The HCES will apply to utilities in the aggregate; utilities may trade compliance credits with other utilities.

- There is no option to purchase compliance credits from the government. All credits are backed by physical generation.
- All electricity retailers are covered by the requirement, regardless of ownership type or size.
- HCES credits earned in one year cannot be "banked" for use in a subsequent year. All credits must be used for compliance in the year that the underlying generation was produced.
- HCES obligations are based on total electricity sales, regardless of source. There is no provision for excluding any electricity sales from a seller's baseline based on resources used to produce the electricity or type of customer purchasing the electricity.
- The HCES operates independent of any state-level policies. The same underlying generation can be used to simultaneously comply with the HCES and any State generation requirements, if otherwise allowed for by both Federal and State law.

Like other EIA analyses of energy and environmental policy proposals, this report focuses on the impacts of those proposals on energy choices in all sectors and the implications of those decisions for emissions and the economy. This focus is consistent with EIA's statutory mission and expertise. The study does not account for any possible health or environmental benefits that might be associated with the HCES policy.

Analysis Cases

The analysis presented in this report starts from the *Annual Energy Outlook 2011 (AEO2011)* Reference case[1] (Ref), which is compared to a case that reflects the HCES requirements outlined in the previous section. The same comparison is repeated under a series of alternative assumptions regarding the costs of generation fuels and technologies. The assumptions used in the eight alternative cases, each of which is run with and without the HCES policy, are briefly summarized below and are more fully explained in Appendix E of the *AEO2011*.

Nuclear Low Cost (LC-Nuc): Capital and operating costs for new nuclear capacity start 20 percent lower than in the Reference case and fall to 40 percent lower in 2035.

Nuclear High Cost (HC-Nuc): Costs for new nuclear technology do not improve from 2011 levels in the Reference case through 2035.

Renewable Low Cost (LC-Ren): Costs of non-hydropower renewable generating technologies start 20 percent lower in 2011 and decline to 40 percent lower than Reference case levels in 2035. Capital costs of renewable liquid fuel technologies start 20 percent lower in 2011 and decline to approximately 40 percent lower than Reference case levels in 2035.

Renewable High Cost (HC-Ren): Costs of non-hydropower renewable generating technologies remain constant at 2011 levels through 2035. Costs are still tied to key commodity price indexes, but no cost improvement from "learning-by-doing" effects is assumed.

Natural Gas Low Cost (LC-Gas) (corresponds with High Shale Recovery case in the AEO2011): The estimated undeveloped technically recoverable shale gas resource base is 50 percent higher than in the Reference case with the per well recovery rate unchanged from the Reference case, resulting in more wells needed to fully recover the resource.

Natural Gas High Cost (HC-Gas) (corresponds with Low Shale Recovery case in the AEO2011): The estimated undeveloped technically recoverable shale gas resource base is 50 percent lower than in the Reference case with the per well recovery rate unchanged from the Reference case, resulting in fewer wells needed to fully recover the resource.

Coal Low Cost (LC-Coal): Regional productivity growth rates for coal mining are approximately 2.7 percent per year higher than in the Reference case, and coal mining wages, mine equipment costs, and coal transportation rates are between 22 and 25 percent lower by 2035 than in the Reference case.

Coal High Cost (HC-Coal): Regional productivity growth rates for coal mining are approximately 2.7 percent per year lower than in the Reference case, and coal mining wages, mine equipment costs, and coal transportation rates are between 25 and 28 percent higher by 2035 than in the Reference case.

RESULTS

HCES Impacts under AEO2011 Reference Case

The HCES results in a large shift in the generation mix (Figure 1 and Table B1). Coal-fired generation, which grows by nearly 23 percent between 2009 and 2035 in the Reference case, decreases by 46 percent between 2009 and 2035 in the HCES case. Coal is primarily displaced by increased natural gas generation, which in the HCES case is 38 percent greater than the Reference case level in 2025 and 30 percent greater in 2035. Nuclear and

billion kilowatthours

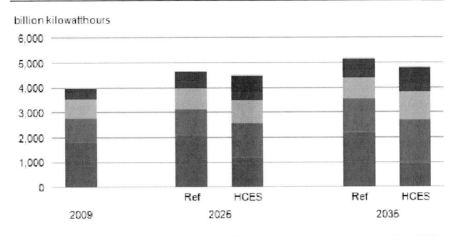

Source: U.S. Energy Information Administration. Natioal Energy Modelling System.runs reghall.do82611b and ceshallnb.d083011a.

Figure 1. Total Net Electricity Generation.

renewable generation also exceed the Reference case projection in the HCES case, though the HCES effect on nuclear generation occurs primarily after 2025.

Among renewable sources, wind and biomass have the largest generation increases under the HCES (Figure 2 and Table B1). By 2035, there is nearly twice as much wind generation than without the HCES policy. Additional biomass generation is met primarily through increased co-firing of biomass in existing coal plants, which decreases in the latter part of the projection as new nuclear generation capacity comes online and existing coal capacity is retired.

HCES compliance strategies vary over time. Compliance through 2020 is attained primarily from existing nuclear and renewable capacity, renewable capacity projected to be built with or without the HCES policy, increasing dispatch of existing qualified natural gas plants, and increasing co-firing of biomass. After 2020, an increasing amount of incremental credits are achieved by generation from wind and nuclear capacity additions in excess of the Reference case, as well as coal-firedgeneration from existing plants retrofitted with sequestration technology.

Annual electricity sector carbon dioxide emissions decrease by more than 50 percent between 2009 and 2035 under the HCES (Figure 3 and Table B1). In the Reference case scenario, however, electricity-sector carbon dioxide emissions increase over the forecast period to reach 2,500 million metric tons of carbon dioxide (MMTCO2) by 2035. In 2025, the electric power sector accounts for 1,525 MMTCO2 under the HCES, which is 35 percent less than in the Reference case. By 2035, HCES electric power sector emissions are 60 percent below the Reference case.

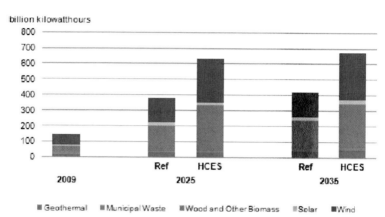

Source: U.S. Energy Information Administration. Natioal Energy Modelling System.runs reghall.do82611b and ceshallnb.d083011a.

Figure 2. Total Non-Hydroelectric Renewable Generation.

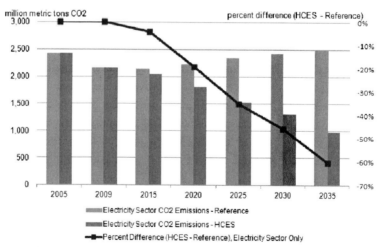

Source: U.S. Energy Information Administration. Natioal Energy Modelling System.runs reghall.do82611b and ceshallnb.d083011a.

Figure 3. Electricity Sector Carbon Dioxide Emissions.

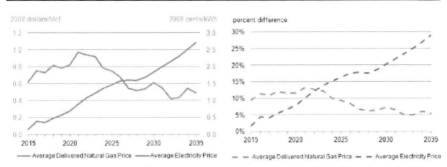

Source: U.S. Energy Information Administration. Natioal Energy Modelling System.runs reghall.do82611b and ceshallnb.d083011a.

Figure 4. HCES Impact on Electricity and Natural Gas Prices (HCES Difference from Reference Case).

The HCES has an increasing impact on average electricity prices from 2015 through 2035 (Figure 4 and Table B1). The impacts on electricity prices prior to 2015 are negligible, because the Reference case projects sufficient eligible generation to nearly meet the HCES requirement. Beyond 2015, electricity prices under the HCES rise above the Reference case level, and the difference grows steadily through 2035. In 2025, the average HCES electricity price is 10.5 cents/kWh – or about 1.5 cents (16 percent) greater than without the policy. In 2035, the average electricity price under the HCES exceeds the Reference case average price by 2.7 cents/kWh (29 percent).

The HCES impact on electricity prices varies significantly across regions (Table 1). In 2035, the HCES impact on average electricity prices ranges between negative 1.6 cents/kWh (indicating that the average electricity price is actually lower under the HCES than the reference case) and positive 8.4 cents/kWh. Regions that are more dependent on generation fuels that are not HCES-eligible, primarily coal, in general experience a stronger price impact.

Natural gas prices increase under the HCES, particularly in the earlier part of the projection. Average delivered natural gas prices exceed Reference case average delivered prices by $0.75/Mcf (9.3 percent) in 2025, but only $0.49/Mcf (5.4 percent) in 2035. Unlike in the case of electricity, the HCES impact on natural gas prices does not increase throughout the entire projection. In earlier years of the legislation, natural gas accounts for much of the incremental HCES compliance, which results in a surge in natural gas prices. As other compliance options are built, however, the differential between natural gas prices with and without the HCES remains between about 5 percent and 10 percent from 2025 to 2035.

Table 1. Regional Electricity Prices (cents/kWh)

	Region	2009	2025 Reference	2025 HCES	2035 Reference	2035 HCES
1	ERCOT - ERCOT All	10.4	9.2	11.8	10.0	14.2
2	FRCC - FRCC All	11.6	10.9	13.4	11.2	16.0
3	MORE - MRO East	9.3	7.5	8.2	7.3	5.6
4	MROW - MRO West	7.6	6.8	8.3	6.9	9.0
5	NEWE - NPCC New England	15.7	13.6	15.0	13.1	16.8
6	NYCW - NPCC NYC/Westchester	19.9	16.8	19.1	16.9	22.3
7	NYLI - NPCC Long Island	18.1	16.7	21.2	16.6	25.1
8	NYUP - NPCC Upstate NY	11.6	11.9	14.1	12.6	17.1
9	RFCE - RFC East	12.2	10.7	13.3	10.9	16.4
10	RFCM - RFC Michigan	9.6	8.7	10.3	9.0	12.2
11	RFCW - RFC West	8.6	8.5	10.9	9.9	12.9
12	SRDA - SERC Delta	7.5	7.3	6.9	7.5	7.3
13	SRGW - SERC Gateway	7.8	6.5	8.5	7.0	11.3
14	SRSE - SERC Southeastern	9.1	8.7	9.2	8.5	9.9
15	SRCE - SERC Central	7.8	6.0	7.0	6.0	8.8
16	SRVC - SERC VACAR	8.6	8.1	8.7	8.3	9.8
17	SPNO - SPP North	7.9	7.6	9.5	7.5	10.2
18	SPNO - SPP South	6.9	7.8	10.2	8.5	12.4
19	AZNM - WECC Southwest	9.8	9.5	10.6	10.4	11.9
20	CAMX - WECC California	13.3	14.6	13.6	13.2	13.8
21	NWPP - WECC Northwest	7.0	4.6	4.6	5.2	5.6
22	RMPA - WECC Rockies	8.2	9.0	12.4	9.4	13.9
	U.S. Average	9.8	9.0	10.5	9.4	12.1

HCES electricity price is 10-25 percent greater than the Reference case electricity price
HCES electricity price is 25 percent or more greater than the Reference case electricity price

Source: U.S. Energy Information Administration. Natioal Energy Modelling System.runs reghall.do82611b and ceshallnb.d083011a.

Note: See Appendix C for a map of the NEMS electricity marketmodule regions.

Electricity expenditures increase under the HCES as a result of higher electricity prices (Figure 5 and Table B1). However, because electricity sales decrease slightly, the impact is smaller than the impact on electricity prices. In 2035, total electricity expenditures under the HCES policy are 18 percent above the projected Reference case level. In 2025, the average household spends $1,277 per year on electricity – $115 above the Reference case –and by 2035, expenditures rise to $1,407 per year – $211 above the Reference case.

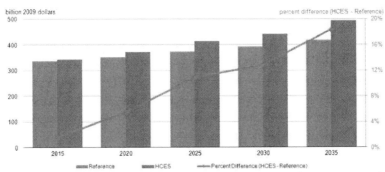

Source: U.S. Energy Information Administration. Natioal Energy Modelling
System.runs reghall.do82611b and ceshallnb.d083011a.

Figure 5. Total Electricity Expenditures.

Higher natural gas prices lead to increased natural gas expenditures
outside the electricity sector under the HCES (Figure 6 and Table B1). In
2025, non-electric natural gas expenditures under the HCES exceed Reference
case expenditures by 8 percent. This differential increases to 10 percent by
2035. In comparison to non-electric natural gas expenditures, natural gas
expenditures in the electric power sector experience a dual upward pressure,
from both higher prices and higher consumption. Particularly in early years,
when increasing natural gas use at existing plants accounts for the greatest
share of HCES compliance, the expenditure effect is quite large.

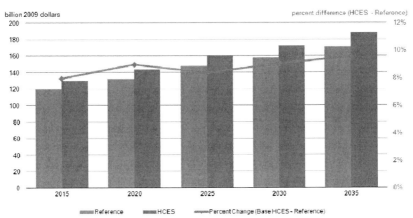

Source: U.S. Energy Information Administration. Natioal Energy Modelling
System.runs reghall.do82611b and ceshallnb.d083011a.

Figure 6. Natural Gas Expenditures, Not Including the Electric Power Sector.

The HCES reduces real GDP relative to the Reference case, though this effect moderates toward the end of the projection period (Figures 7 and 8 and Table B1). The peak negative impact is less than eight-tenths of one percent, realized in 2024. In the latter part of the projection, however, GDP under the HCES converges back toward the Reference case. GDP grows at an average annual rate of 2.68 percent between 2009 and 2035 under the HCES, just slightly below the Reference case growth rate of 2.69 percent. Real GDP per capita[2] in 2035 is $65,658 under the HCES, versus $65,848 in the Reference case – a reduction of about 0.3 percent.

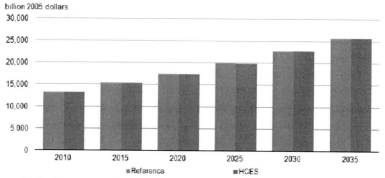

Source: U.S. Energy Information Administration. Natioal Energy Modelling System.runs reghall.do82611b and ceshallnb.d083011a.

Figure 7. Annual GDP.

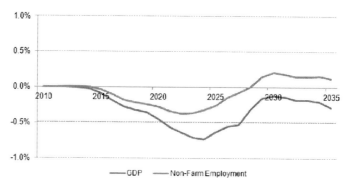

Source: U.S. Energy Information Administration. Natioal Energy Modelling System.runs reghall.do82611b and ceshallnb.d083011a.

Figure 8. HCES Impact on Employment and Real GDP, Percent Difference (HCES Difference from Reference Case).

The HCES negatively affects non-farm employment from 2015 through the mid-2020's, but employment recovers toward the end of the projection period, following the trend of GDP. The change in overall energy prices peaks in 2025 and then begins to return to Reference case levels. In addition, the amount of diverted energy investment peaks in the mid-2020's, resulting in fewer diverted resources and productivity impacts later in the projection period. Service-sector employment leads the employment recovery, as services use relatively less energy than the manufacturing sector.

Sensitivity Analysis

The HCES could have a different effect when resource or technology costs diverge from the assumptions used in the Reference case. The following section considers the effect of the HCES when applied to different baseline scenarios. Per the request from Chairman Hall, EIA models the effect of the HCES given nine sensitivity scenarios, each of which are described in the introduction to this report.[3] Therefore, this section considers eighteen individual model scenarios – nine baseline sensitivity scenarios, and then the HCES under each of those scenarios. For the purpose of presenting the material in a digestible format, most of the discussion and Figures 10, 11, 12, and 14 below focus on the *impact* of the HCES, which is always described in reference to a specific corresponding baseline scenario. For example, the impact of the HCES on electricity prices in the low-cost nuclear case compares electricity prices under the HCES in the low-cost nuclear scenario to electricity prices in the low-cost nuclear case without the HCES. This approach isolates the effect of the policy from the underlying scenario assumptions. For this reason, the HCES cases with the highest or lowest impact on a given indicator do not necessarily reflect the cases that yield the highest or lowest level of that indicator. Tables B2 through B5 provide results for levels in all of the sensitivity cases.

The HCES causes coal-based generation to decline significantly in all sensitivity cases (Figure 9). In 2009, coal plants provided 45 percent of total power generation. However, by 2025 the share of generation from coal ranges from 22 percent to 27 percent in the HCES sensitivity cases, versus 41 percent to 46 percent in the base cases. The fall continues after 2025, when the share ranges from 10 percent to 20 percent in 2035 in the HCES sensitivity cases, versus 37 percent to 44 percent in the base cases. Of the HCES sensitivity cases, the highest share for coal occurs in the high-cost natural gas HCES case,

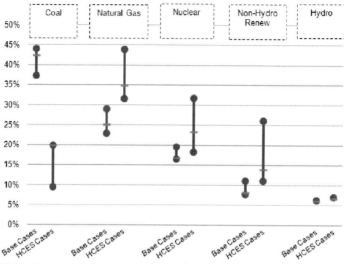

Source: U.S. Energy Information Administration. Natioal Energy Modelling System.runs refhall.d082611b, ceshallnb.d083011a, refhallhn.d082611b, cashallnbhn.d083011a, refhallhn.d082611b, ceshallnb.d083011a, refhallhn.d082611b, ceshallnb.d083011a, refhallhn.d082611b, ceshallnb.d083011a, refhallhn.d082611b, ceshallnb.d083011a, refhallhn.d082611b, ceshallnb.d083011a, refhallhn.d082611b, ceshallnb.d083011a.

Note: The green dash shows the reference case value on the "base cases" graphs, and the basic HCES case (ie HCES applied to the AE02011 Reference case) on the "CES cases" graphs.

Figure 9. Fuel Shares of Total Generation in 2035, Range Over Sensitivity Cases.

while the lowest occurs in the high-cost coal HCES case. The HCES has the greatest impact – or causes the greatest reduction in coal-fired generation – in the low-cost renewable sensitivity case.

In contrast to the situation for coal, natural gas generation and non-hydroelectric renewable generation each increase significantly in the HCES sensitivity cases. However, there is significant variation in their share of total generation, depending on the underlying assumptions about their costs and the costs of other technologies. The share of generation coming from natural gas in the HCES sensitivity cases in 2035 varies from 32 percent to 44 percent, compared to 23 percent to 29 percent in the base cases. Among the HCES sensitivity cases, the highest share for natural gas occurs in the high-cost coal HCES case, while the lowest share occurs in the low-cost nuclear HCES case.

Natural gas generation is most significantly impacted by the HCES in the high-cost nuclear case, where natural gas generation under the HCES exceeds the base case by 51 percent. The share of generation coming from non-hydroelectric renewables in the HCES sensitivity cases in 2035 varies from 11 percent to 26 percent – again, well above the 8 percent to 11 percent range of the base cases. The highest share occurs in the low-cost renewable HCES case and the lowest shares occur in the high-cost renewables and low-cost nuclear HCES cases. However, the impact of the HCES on the non-hydroelectric renewable generation is greatest in the low-cost renewable sensitivity case, in which non-hydroelectric renewable generation under the HCES exceeds the base case level by 118 percent.

Nuclear generation also increases under the HCES relative to baseline scenarios. However, the magnitude of the effect is extremely sensitive to the underlying baseline scenario. In the high-cost nuclear scenario, nuclear generation under the HCES is only 0.8 percent greater in 2035 than the associated low-cost nuclear baseline. In contrast, nuclear generation under low-cost nuclear assumptions with the HCES exceeds the low-cost nuclear baseline by 54.6 percent. Significant effects on nuclear generation are primarily concentrated in the latter part of the projection period (2025 and after).

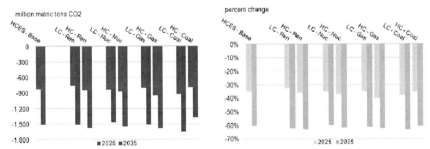

Source: U.S. Energy Information Administration. Natioal Energy Modelling System.runs refhall.d082611b, ceshallnb.d083011a, refhallhn.d082611b, cashallnbhn.d083011a, refhallhn.d082611b, ceshallnb.d083011a, refhallhn.d082611b, ceshallnb.d083011a, refhallhn.d082611b, ceshallnb.d083011a, refhallhn.d082611b, ceshallnb.d083011a, refhallhn.d082611b, ceshallnb.d083011a, ceshallnb.d083011a.

Figure 10. HCES Impact on Carbon Dioxide Emissions (HCES Difference from Corresponding Base Case).

Natural gas is the leading source of generation by 2035 under the HCES in most of the HCES sensitivity cases. The notable exception to this trend is in the low-cost nuclear scenario, where relatively affordable nuclear capacity displaces natural gas as HCES-qualified baseload generation.

Carbon dioxide emissions in the electric power sector fall significantly as a result of the HCES in all sensitivity cases (Figure 10). In each sensitivity case, the HCES results in emissions that are 33 percent to 40 percent lower than the associated base case levels in 2025, and 60 percent to 64 percent lower than the associated base case levels in 2035. Reductions are most significant in the low-cost coal scenario.

Conversely, reductions in the high-cost coal scenario appear to be relatively modest – however, this is somewhat misleading, because the absolute level of emissions is actually lowest in the high-cost coal sensitivity case. The high cost of coal drives a reduction in coal-fired generation regardless of the HCES policy, and, therefore, the HCES policy has a lesser impact.

The HCES policy leads to higher electricity prices in all of the sensitivity cases (Figure 11). All alternative side cases exhibit higher average electricity prices under the HCES compared to the corresponding baseline. For example, the average electricity price in the baseline low-cost nuclear scenario is 9.3 cents/kWh in 2035, but with the HCES policy, the price is 11.0 cents/kWh. The difference between HCES and baseline electricity prices ranges from 1.7 cents/kWh to 3.6 cents/kWh in 2035. Electricity prices in 2035 without the HCES range from 8.9 cents/kWh to 10.0 cents/kWh, while under the HCES they range from 11.0 cents/kWh to 13.2 cents/kWh. Total and average household electricity expenditures follow a similar pattern, increasing across various sensitivity cases with the HCES. However, the price effect is again dampened by the resultant reduction in electricity sales, which ranges from 3.9 percent to 6.9 percent in the residential sector. The impact of the HCES on average household electricity expenditures ranges from increases of $131 to $279 per year in 2035 – or 11 percent to 23 percent above baseline expenditures.

Electricity prices under the high-cost renewables scenario exhibit greater sensitivity to the HCES than in the other cases. Conversely, the price impact of the HCES is lowest in the low-cost nuclear scenario.

In the high-cost renewables scenario, utilities still install significantly more non-hydroelectric renewable electricity than in the baseline high-cost renewable scenario. Because this technology is relatively more expensive to build, this additional cost translates into higher HCES credit prices (that is, compliance costs), which, in turn, increases electricity prices. In the low-cost nuclear scenario, the HCES has a relatively minimal impact over time, because a larger portion of overall HCES compliance can be met through generation from new nuclear capacity, the cost of which this scenario sets to be 40 percent less than the Reference case in 2035.

Natural gas prices generally increase under the HCES; however, the magnitude of this impact decreases toward the end of the projection horizon as other compliance options are increasingly available and attractive (Figure 12). This temporal pattern is generally consistent when the HCES is applied to alternative baseline scenarios.

Interestingly, in the low-cost nuclear scenario, natural gas prices under the HCES in 2035 are actually lower than without the HCES policy, due to the much greater amount of nuclear generation capacity that is built in the latter part of this scenario. The HCES has the greatest price impact on natural gas in the high-cost natural gas case.

The finding that the HCES results in lower GDP is also robust across scenarios. However, consistent with the main case results, the impact on the growth rate of GDP is small. The average annual GDP growth rate over the

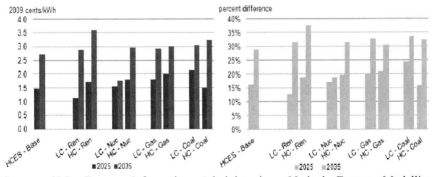

Source: U.S. Energy Information Administration. Natioal Energy Modelling System.runs refhall.d082611b, ceshallnb.d083011a, refhallhn.d082611b, cashallnbhn.d083011a, refhallhn.d082611b, ceshallnb.d083011a, refhallhn.d082611b, ceshallnb.d083011a, refhallhn.d082611b, ceshallnb.d083011a, refhallhn.d082611b, ceshallnb.d083011a, refhallhn.d082611b, ceshallnb.d083011a, refhallhn.d082611b, ceshallnb.d083011a, refhallhn.d082611b, ceshallnb.d083011a.

Figure 11. HCES Impact on Electricity Prices (HCES Difference from Corresponding Base Case).

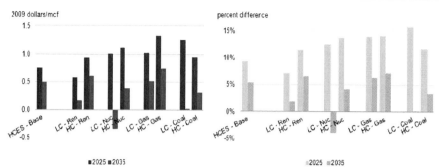

Source: U.S. Energy Information Administration. Natioal Energy Modelling System.runs refhall.d082611b, ceshallnb.d083011a, refhallhn.d082611b, cashallnbhn.d083011a, refhallhn.d082611b, ceshallnb.d083011a, refhallhn.d082611b, ceshallnb.d083011a, refhallhn.d082611b, ceshallnb.d083011a, refhallhn.d082611b, ceshallnb.d083011a, refhallhn.d082611b, ceshallnb.d083011a.

Figure 12. HCES Impact on Delivered Natural Gas Prices (HCES Difference from Corresponding Base Case).

2009 to 2035 period ranges from 2.66 percent to 2.69 percent across the range of HCES sensitivity cases, compared to 2.68 percent to 2.69 percent in the corresponding base cases. In 2035, annual GDP ranges from $25,623 billion to $25,710 billion in the base case scenarios, versus a range of $25,514 billion to $25,705 billion under the HCES legislation (Figure 13). On a per capita basis, this translates to base case ranges between $65,686 per person and $65,909 per person, compared to a range of $65,406 per person to $65,897 per person under the HCES.

The negative effect on cumulative discounted GDP between 2009 and 2035 is less than 0.3 percent in all scenarios (Figure 14). In most sensitivity cases, annual GDP exhibits a recovery relative to the corresponding base case in the latter part of the projection (recall Figure 8).

The nearer-term (2025) impact is strongest in the low-cost gas, high-cost nuclear, and low-cost coal scenarios. In the latter case, the differential is large because utilities cannot fully take advantage of the low-cost coal while still complying with the HCES. This forces retirement of plants that would be able to produce electricity relatively cheaply, and diverts investment from lower cost alternatives.

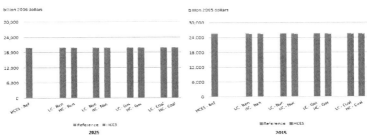

Source: U.S. Energy Information Administration. Natioal Energy Modelling System.runs refhall.d082611b, ceshallnb.d083011a, refhallhn.d082611b, cashallnbhn.d083011a, refhallhn.d082611b, ceshallnb.d083011a, refhallhn.d082611b, ceshallnb.d083011a, refhallhn.d082611b, ceshallnb.d083011a, ceshallnb.d083011a, refhallhn.d082611b, ceshallnb.d083011a, refhallhn.d082611b, ceshallnb.d083011a.

Figure 13. Annual GDP.

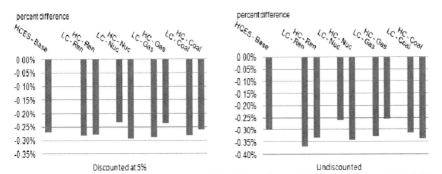

Source: U.S. Energy Information Administration. Natioal Energy Modelling System.runs refhall.d082611b, ceshallnb.d083011a, refhallhn.d082611b, cashallnbhn.d083011a, refhallhn.d082611b, ceshallnb.d083011a, refhallhn.d082611b, ceshallnb.d083011a, refhallhn.d082611b, ceshallnb.d083011a, refhallhn.d082611b, ceshallnb.d083011a, ceshallnb.d083011a.

Figure 14. HCES Impact on Cumulative (2009-2035) GDP (HCES Difference from Corresponding Base Case).

APPENDIX A. REQUEST LETTER

RALPH M. HALL, TEXAS
CHAIRMAN

EDDIE BERNICE JOHNSON, TEXAS
RANKING MEMBER

U.S. HOUSE OF REPRESENTATIVES

COMMITTEE ON SCIENCE, SPACE, AND TECHNOLOGY

2321 RAYBURN HOUSE OFFICE BUILDING
WASHINGTON, DC 20515-6301
(202) 225-6371
www.science.house.gov

July 22, 2011

The Honorable Howard Gruenspecht
Acting Administrator
Energy Information Administration
U.S. Department of Energy
1000 Independence Avenue, SW
Washington, DC 20585

Dear Administrator Gruenspecht:

On March, 15, 2011, I wrote then-Administrator Newell requesting an Energy Information Administration (EIA) analysis of the economic impacts of a Clean Energy Standard (CES). The purpose of this letter is to more fully define the assumptions for that study and to recommend the specific analyses I would like you to undertake.

The attached document details this request, which was developed after consultation with your staff. In brief, I request that you estimate the impact of the proposed CES on seven different economic factors, beginning with the base policy scenario as defined by the Annual Energy Outlook 2011 (AEO2011) and then modified using nine additional scenarios as defined in the attachment.

Should you have any further questions, please contact Andy Zach, Professional Staff with the Energy and Environment Subcommittee. In advance, thank you for your assistance.

Sincerely,

Ralph M. Hall

Ralph M. Hall
Chairman

cc: Secretary Steven Chu

Attachment: Details of Chairman Hall CES Analysis Request

Because of the uncertainties associated with the structure and legislative details of a CES, we would like the following details incorporated into the "Best Estimate CES" scenario.

- Eligible resources to meet the target will include: hydroelectric, wind, solar, geothermal, biomass power, municipal solid waste, landfill gas, nuclear, coal-fired plants with carbon capture and sequestration, and natural gas-fired plants with either carbon capture and sequestration or utilizing combined cycle technology. Generation may derive from the electric power sector or from industrial, commercial, or residential generators using qualified resources. Qualifying generation will be determined solely by resource and technology, and not by vintage of the plant or by difference from historic generation at a plant.
- CES target would start from an initial share of 40 percent (qualified generation as a percent of sales), utilities will achieve 80 percent qualified generation by 2035. Because the 40 percent is specified from historical values (2010), and the target share is to increase linearly through the ramping period, EIA will assume that the policy has an initial target of 44.8 percent in 2013. The target will increase by 1.6 percentage points each year thereafter, achieving 80 percent by 2035.
- There will be no sunset in the CES requirement. The 80 percent target will remain constant from 2035 onward.
- The "Best Estimate CES" case will assume utilities may trade credits for generation. The CES target will apply to utilities in the aggregate, and some utilities may generate more electricity from eligible resources and may trade compliance credits to other utilities, who may then apply those credits to a compliance deficit.
- Compliance with CES targets will be based on accumulated credits. In general, and unless otherwise indicated, credits will be worth a "face value" of 1 MWh for each MWh of generation. Credits for natural gas fired in a combined cycle will count 50 percent toward compliance (a utility will earn 0.5 MWh of compliance credits for every 1 MWh of natural gas generation from a combined cycle plant.) Credits from coal or natural gas with carbon capture and sequestration will count 90 percent towards compliance.
- There will be no option to purchase compliance credits from the government. All credits must be backed by physical generation.
- All utilities are covered by the requirement, regardless of ownership status or size.
- Utilities would not be able to "bank" excess credits earned in one year to be used for compliance in a subsequent year. All credits must be used for compliance in the year that the underlying generation was produced.
- Generation targets are specified based on sales of all electricity, regardless of source. There is no provision for excluding any electricity sales from each utility's baseline based on resources used to produce the lectricity or type of customer purchasing the electricity.
- The model will assume a national CES does not interfere with any similar policies in effect at the state level. Utilities may use the same underlying generation to simultaneously comply with any State generation requirements, if otherwise allowed for by both Federal and State law.

Utilizing the parameters outlined above, please examine several scenarios. In addition to examining the base policy scenario, as defined by the Annual Energy Outlook 2011 (AEO2011), please outline the following scenarios:

1. Best Estimate CES, as defined above;
2. Low Cost Nuclear, same as Best Estimate CES, but incorporating the "Low Cost Nuclear" assumptions developed for an AEO2011 summary case;
3. High Cost Nuclear, same as Best Estimate CES, but incorporating the "High Cost Nuclear" scenario developed as an AEO2011 summary case;
4. Low Cost Renewable, same as Best Estimate CES, but incorporating the "Low Cost renewable" scenario developed as an AEO2011 summary case;
5. High Cost renewable, same as Best Estimate CES, but incorporating "High Cost Renewable" scenario developed as an AEO2011 summary case;
6. Low Shale Gas Recovery, same as Best Estimate CES, but incorporating the assumptions from the "Low Shale Estimated Ultimate Recovery" case in the AEO2011;
7. High Shale Gas Recovery, same as Best Estimate CES, but incorporating the assumptions from the "High Shale Estimated Ultimate Recovery" case in the AEO2011;
8. High Coal Cost, same as Best Estimate CES, but incorporating the assumptions from the "High Coal Cost" scenario in the AEO2011;
9. Low Coal Cost, same Best Estimate CES, but incorporating the assumptions from the "Low Coal Cost" scenario in the AEO2011.

For each of the scenarios outlined above, please calculate
- projected average cost of electricity generation per megawatt-hour;
- overall nationwide electricity generation costs;
- average cost of electricity per household;
- national gross domestic product;
- gross domestic product per capita; and
- national employment levels.

APPENDIX B. SUMMARY TABLES

Table B1. The HCES compared to the Reference case

	2009	2025		2035	
		Reference	HCES	Reference	HCES
Generation (billion kilowatthours)					
Coal	1,772	2,049	1,156	2,184	951
Petroleum	41	45	44	47	45
Natural Gas	931	1,002	1,386	1,293	1,676
Nuclear	799	871	928	868	1,127
Conventional Hydropower	274	306	320	314	321
Geothermal	15	25	26	42	49
Municipal Waste	18	17	17	17	17
Wood and Other Biomass	38	162	291	181	281
Solar	3	18	18	21	23

	2009	2025		2035	
		Reference	HCES	Reference	HCES
Wind	71	153	277	159	301
Other	18	16	16	16	16
Total Generation	3,981	4,665	4,479	5,142	4,807
Capacity (gigawatts)					
Coal	317	323	262	330	267
Petroleum	116	87	87	87	86
Natural Gas	351	382	384	455	444
Nuclear	101	110	117	110	143
Conventional Hydropower	78	79	82	81	82
Geothermal	2	3	4	6	6
Municipal Waste	4	4	4	4	4
Wood and Other Biomass	7	17	17	20	20
Solar	2	11	11	13	14
Wind	32	53	92	55	100
Other (including pumped storage)	24	25	25	25	25
Total	1,033	1,095	1,086	1,185	1,193
Prices (2009 cents/kWh)					
Credit Price			8.1		10.8
Electricity Price	9.8	9.0	10.5	9.4	12.1
Residential	11.5	10.7	12.2	10.9	13.6
Commercial	10.1	9.3	10.8	9.4	12.2
Industrial	6.8	6.3	7.5	6.6	8.9
Average Delivered Natural Gas Price (2009 dollars/Mcf)	7.5	8.1	8.8	9.2	9.7
Expenditures (billion 2009 dollars except as noted)					
Total Electricity Expenditures	350	373	414	417	494
Residential Electricity Expenditures	156	157	172	176	207
Household Electricity Expenditures (2009 Dollars/Household)	1379	1162	1277	1196	1407
Total Natural Gas Expenditures	156	187	225	227	264
Electricity Sector Natural Gas Expenditures	34	39	65	55	77
Non-Electricity Sector Natural Gas Expenditures	122	148	160	171	188
CES Compliance					
Credits Required (percent of sales)			64		80
Credits Achieved (percent of sales)			64		78
Generation Achieved (percent of sales)			64		78
Total Electricity Sales (billion kilowatthours)	3,556	4,105	3,913	4,428	4,064
Emissions					
Sulfur Dioxide (million metric tons)	5.7	4.1	3.1	3.7	2.4
Nitrogen Oxide (million metric tons)	2.0	2.0	1.5	2.0	1.2
Mercury (metric tons)	41	29	16	29	15
Carbon Dioxide (million metric tons CO_2)	2,160	2,345	1,525	2,500	991
Macro Economic					

Table B1. (Continued)

	2009	2025		2035	
		Reference	HCES	Reference	HCES
GDP (billion 2005 dollars)	12,881	20,012	19,885	25,686	25,612
Per Capita GDP (thousand 2005 dollars/person)	42	56	56	66	66
Employment, Non-Farm (million)	131	156	156	171	171
Employment, Manufacturing (million)	12	16	15	13	13

Source: U.S. Energy Information Administration. National Energy Modeling System, runs refhall.d082611b, ceshallnb.d083011a.

Table B2. Low and high-cost renewable scenarios: The HCES compared to the sensitivity base cases

	2009	2025				2035			
		Low Cost Renewable		high Cost Renewable		Low Cost Renewable		high Cost Renewable	
		Base	HCES	Base	HCES	Base	HCES	Base	HCES
Generation (billion kilowatthours)									
Coal	1,772	2,030	1,238	2,034	1,126	2,142	559	2,134	762
Petroleum	41	45	44	46	43	47	43	48	44
Natural Gas	931	979	1,155	994	1,409	1,192	1,687	1,308	1,917
Nuclear	799	877	877	877	938	874	898	874	1,097
Conventional Hydropower	274	313	324	306	316	326	340	314	321
Geothermal	15	27	34	25	26	44	36	29	27
Municipal Waste	18	17	17	17	17	17	17	17	17
Wood and Other Biomass	38	182	344	150	236	205	341	145	182
Solar	3	24	25	16	16	48	75	17	18
Wind	71	168	478	158	277	261	787	186	287
Other	18	16	16	16	16	16	16	16	16
Total Generation	3,981	4,680	4,552	4,640	4,419	5,173	4,800	5,089	4,689
Capacity (gigawatts)									
Coal	317	322	260	321	261	330	229	327	260
Petroleum	116	87	87	88	86	87	86	86	86
Natural Gas	351	378	375	384	386	439	433	460	454
Nuclear	101	110	110	110	119	110	114	110	139
Conventional Hydropower	78	80	83	79	81	84	88	80	82
Geothermal	2	4	5	3	4	6	5	4	4
Municipal Waste	4	4	4	4	4	4	4	4	4
Wood and Other Biomass	7	18	24	11	11	22	38	12	12
Solar	2	15	15	10	10	27	41	11	11
Wind	32	58	165	55	91	88	277	64	95
Other (including pumped storage)	24	25	25	25	25	25	25	25	25
Total	1,033	1,101	1,153	1,090	1,077	1,222	1,339	1,183	1,171

	2009	2025 Low Cost Renewable		2025 high Cost Renewable		2035 Low Cost Renewable		2035 high Cost Renewable	
		Base	HCES	Base	HCES	Base	HCES	Base	HCES
Prices (2009 cents/kWh)									
Credit Price			6.6		8.6		12.4		14.0
Electricity Price	9.8	8.9	10.0	9.1	10.9	9.1	11.9	9.5	13.1
Residential	11.5	10.6	11.7	10.8	12.5	10.6	13.4	11.1	14.6
Commercial	10.1	9.0	10.2	9.4	11.2	9.1	12.1	9.6	13.3
Industrial	6.8	6.1	7.1	6.3	7.8	6.4	8.8	6.7	9.9
Average Delivered Natural Gas Price (2009 dollars/MCF)	7.5	8.0	8.6	8.1	9.1	8.9	9.1	9.3	9.9
Expenditures (billion 2009 dollars except as noted)									
Total Electricity Expenditures	350	366	398	377	423	406	482	423	524
Residential Electricity Expenditures	156	154	166	158	175	171	201	178	219
Household Electricity Expenditures (2009 Dollars/Household)	1,379	1,143	1,231	1,173	1,303	1,162	1,369	1,210	1,489
Total Natural Gas Expenditures	156	185	206	189	232	216	251	230	288
Electricity Sector Natural Gas Expenditures	34	38	50	39	68	49	72	57	92
Non-Electricity Sector Natural Gas Expenditures	122	147	156	149	165	167	179	174	195
CES Compliance									
Credits Required (percent of sales)			64		64		80		80
Credits Achieved (percent of sales)			64		63		79		78
Generation Achieved (percent of sales)			64		63		79		78
Total Electricity Sales (billion kilowatthours)	3,556	4,112	3,961	4,101	3,876	4,446	4,016	4,416	3,971
Emissions									
Sulfur Dioxide (million metric tons)	5.7	4.1	3.3	4.2	3.1	3.8	1.3	3.8	1.8
Nitrogen Oxide (million metric tons)	2.0	2.0	1.5	2.0	1.4	2.0	0.7	2.0	1.0
Mercury (metric tons)	41	29	16	29	15	29	7	28	12
Carbon Dioxide (million metric tons CO2)	2,160	2,318	1,563	2,333	1,491	2,421	914	2,475	914
Macro Economic									
GDP (billion 2005 dollars)	12,881	20,019	19,930	19,988	19,861	25,703	25,595	25,674	25,521
Per Capita GDP (thousand 2005 dollars/person)	42	56	56	56	55	66	66	66	65
Employment, Non-Farm (million)	131	156	156	156	155	171	171	171	170

Table B2. (Continued)

	2009			2025			2035		
		Low Cost Renewable		high Cost Renewable		Low Cost Renewable		high Cost Renewable	
		Base	HCES	Base	HCES	Base	HCES	Base	HCES
Employment, Manufacturing (million)	12	16	16	16	15	13	13	13	13

Sources: U.S. Energy Information Administration. National Energy Modeling System, runs refhall.d082611b, ceshallnb.d083011a, refhallhc.d082611b, ceshallnbhr.d083011a, refhalllr.d082611b, ceshallnblr.d083011a.

Table B3. Low and high-cost nuclear scenarios: The HCES compared to the sensitivity base cases

	2009			2025			2035		
		Low Cost Nuclear		high Cost Nuclear		Low Cost Nuclear		high Cost Nuclear	
		Base	HCES	Base	HCES	Base	HCES	Base	HCES
Generation (billion kilowatthours)									
Coal	1,772	2,047	1,110	2,060	1,062	2,169	897	2,185	838
Petroleum	41	45	43	45	44	47	44	46	45
Natural Gas	931	999	1,417	996	1,486	1,184	1,559	1,290	1,943
Nuclear	799	877	1,023	871	877	1,012	1,564	868	874
Conventional Hydropower	274	305	315	305	315	312	315	314	322
Geothermal	15	24	26	25	29	39	41	43	49
Municipal Waste	18	17	17	17	17	17	17	17	17
Wood and Other Biomass	38	162	283	159	284	183	265	178	265
Solar	3	18	18	18	18	21	23	21	26
Wind	71	154	180	154	280	158	198	161	391
Other	18	16	16	16	16	16	16	16	16
Total Generation	3,981	4,666	4,449	4,667	4,431	5,159	4,940	5,140	4,789
Capacity (gigawatts)									
Coal	317	322	260	322	265	330	257	330	265
Petroleum	116	87	88	87	87	87	85	86	84
Natural Gas	351	381	382	383	385	438	423	457	471
Nuclear	101	110	130	110	110	128	200	110	110
Conventional Hydropower	78	78	81	78	81	80	81	81	83
Geothermal	2	3	4	3	4	5	5	6	6
Municipal Waste	4	4	4	4	4	4	4	4	4
Wood and Other Biomass	7	17	17	17	17	20	20	20	21
Solar	2	11	11	11	11	13	14	13	15
Wind	32	54	61	54	94	55	67	56	131
Other (including pumped storage)	24	25	25	25	25	25	25	25	25
Total	1,033	1,093	1,062	1,095	1,083	1,185	1,181	1,187	1,215

	2009	2025 Low Cost Nuclear		2025 high Cost Nuclear		2035 Low Cost Nuclear		2035 high Cost Nuclear	
		Base	HCES	Base	HCES	Base	HCES	Base	HCES
Prices (2009 cents/kWh)									
Credit Price			9.2		9.7		8.5		12.4
Electricity Price	9.8	9.0	10.6	9.0	10.8	9.3	11.0	9.4	12.4
Residential	11.5	10.7	12.2	10.7	12.5	10.8	12.5	10.9	13.9
Commercial	10.1	9.3	10.9	9.2	11.1	9.3	11.1	9.5	12.5
Industrial	6.8	6.3	7.6	6.2	7.8	6.5	8.0	6.6	9.1
Average Delivered Natural Gas Price (2009 dollars/MCF)	7.5	8.1	9.1	8.0	9.1	9.0	8.6	9.1	9.5
Expenditures (billion 2009 dollars except as noted)									
Total Electricity Expenditures	350	373	416	372	422	414	461	419	502
Residential Electricity Expenditures	156	156	172	156	175	174	194	176	210
Household Electricity Expenditures (2009 Dollars/Household)	1,379	1,159	1,277	1,160	1,298	1,186	1,317	1,199	1,431
Natural Gas Expenditures	156	187	233	187	239	217	232	226	279
Electricity Sector Natural Gas Expenditures	34	39	69	39	74	48	61	55	92
Non-Electricity Sector Natural Gas Expenditures	122	148	164	148	165	168	171	171	187
CES Compliance									
Credits Required (percent of sales)			64		64		80		80
Credits Achieved (percent of sales)			63		64		80		79
Generation Achieved (percent of sales)			63		64		80		79
Total Electricity Sales (billion kilowatthours)	3,556	4,105	3,907	4,106	3,886	4,441	4,168	4,424	4,030
Emissions									
Sulfur Dioxide (million metric tons)	5.7	4.2	2.9	4.3	3.0	3.8	2.1	3.9	2.0
Nitrogen Oxide (million metric tons)	2.0	2.0	1.5	2.0	1.4	2.0	1.1	2.0	1.0
Mercury (metric tons)	41	29	15	29	16	29	14	30	13
Carbon Dioxide (million metric tons CO2)	2,160	2,342	1,511	2,352	1,477	2,447	978	2,498	947
Macro Economic									
GDP (billion 2005 dollars)	12,881	20,011	19,862	20,012	19,860	25,708	25,705	25,684	25,588
Per Capita GDP (thousand 2005 dollars/person)	42	56	55	56	55	66	66	66	66
Employment, Non-Farm (million)	131	156	156	156	156	171	171	171	171

Table B3. (Continued)

	2009			2025				2035	
		Low Cost Nuclear		high Cost Nuclear		Low Cost Nuclear		high Cost Nuclear	
	Base	HCES		Base	HCES	Base	HCES	Base	HCES
Employment, Manufacturing (million)	12	16	15	16	15	13	13	13	13

Sources: U.S. Energy Information Administration. National Energy Modeling System, runs refhall.d082611b, ceshallnb.d083011a, refhallhn.d082611b, ceshallnbhn.d083011a, refhallln.d082611b, ceshallnbln.d083011a.

Table B4. Low and high-cost natural gas scenarios: The HCES compared to the sensitivity base cases

	2009			2025				2035	
		Low Cost Natural Gas		high Cost Natural Gas		Low Cost Natural Gas		high Cost Natural Gas	
	Base	HCES		Base	HCES	Base	HCES	Base	HCES
Generation (billion kilowatthours)									
Coal	1,772	1,948	987	2,134	1,078	2,078	771	2,239	941
Petroleum	41	46	45	45	43	47	45	48	44
Natural Gas	931	1,138	1,674	856	1,304	1,475	1,996	1,166	1,503
Nuclear	799	862	910	877	970	860	1,074	874	1,210
Conventional Hydropower	274	305	321	308	315	312	323	314	322
Geothermal	15	25	29	27	29	39	50	44	48
Municipal Waste	18	17	17	17	17	17	17	17	17
Wood and Other Biomass	38	168	256	155	289	184	248	165	283
Solar	3	18	18	18	19	21	23	22	25
Wind	71	145	186	161	291	152	280	180	319
Other	18	16	16	16	16	16	16	16	16
Total Generation	3,981	4,690	4,461	4,616	4,371	5,201	4,844	5,086	4,730
Capacity (gigawatts)									
Coal	317	314	257	327	271	321	256	336	273
Petroleum	116	93	87	86	84	93	86	86	84
Natural Gas	351	386	394	375	368	468	469	440	421
Nuclear	101	109	115	110	123	109	136	110	154
Conventional Hydropower	78	78	82	79	81	80	83	80	83
Geothermal	2	3	4	4	4	5	6	6	6
Municipal Waste	4	4	4	4	4	4	4	4	4
Wood and Other Biomass	7	17	17	17	17	20	20	20	20
Solar	2	11	11	11	12	12	14	13	15
Wind	32	51	63	56	97	53	92	62	106

	2009	2025				2035			
		Low Cost Natural Gas		high Cost Natural Gas		Low Cost Natural Gas		high Cost Natural Gas	
Other (including pumped storage)	24	25	25	25	25	25	25	25	25
Total	1,033	1,091	1,060	1,095	1,086	1,191	1,191	1,182	1,191
Prices (2009 cents/kWh)									
Credit Price			9.2		13.1		11.0		13.6
Electricity Price	9.8	8.8	10.6	9.4	11.4	8.9	11.9	9.8	12.8
Residential	11.5	10.6	12.3	11.1	13.0	10.5	13.4	11.3	14.3
Commercial	10.1	9.0	10.9	9.7	11.8	8.9	11.9	10.0	13.1
Industrial	6.8	6.1	7.6	6.6	8.3	6.2	8.7	7.0	9.6
Average Delivered Natural Gas Price (2009 dollars/MCF)	7.5	7.3	8.3	9.4	10.8	8.1	8.7	10.3	11.1
Expenditures (billion 2009 dollars except as noted)									
Total Electricity Expenditures	350	366	415	386	444	401	484	436	519
Residential Electricity Expenditures	156	155	173	161	181	171	204	181	214
Household Electricity Expenditures (2009 Dollars/Household)	1,379	1,147	1,285	1,192	1,346	1,164	1,391	1,232	1,458
Natural Gas Expenditures	156	179	232	205	264	215	259	243	283
Electricity Sector Natural Gas Expenditures	34	41	78	40	76	59	80	55	83
Non-Electricity Sector Natural Gas Expenditures	122	138	153	166	188	156	179	189	201
CES Compliance									
Credits Required (percent of sales)			64		64		80		80
Credits Achieved (percent of sales)			63		65		79		79
Generation Achieved (percent of sales)			63		65		79		79
Total Electricity Sales (billion kilowatthours)	3,556	4,112	3,880	4,081	3,869	4,460	4,061	4,408	4,022
Emissions									
Sulfur Dioxide (million metric tons)	5.7	3.9	2.8	4.0	3.0	3.8	1.8	3.7	2.2
Nitrogen Oxide (million metric tons)	2.0	2.0	1.4	2.0	1.4	2.0	1.0	2.1	1.1
Mercury (metric tons)	41	27	15	29	15	27	12	30	15

Table B4. (Continued)

	2009	Low Cost Natural Gas		2025 high Cost Natural Gas		Low Cost Natural Gas		2035 high Cost Natural Gas	
Carbon Dioxide (million metric tons CO_2)	2,160	2,290	1,487	2,387	1,434	2,450	945	2,527	948
Macro Economic									
GDP (billion 2005 dollars)	12,881	20,030	19,835	19,962	19,846	25,704	25,643	25,677	25,573
Per Capita GDP (thousand 2005 dollars/person)	42	56	55	56	55	66	66	66	66
Employment, Non-Farm (million)	131	156	155	156	156	171	171	171	171
Employment, Manufacturing (million)	12	16	15	16	15	13	13	13	13

Sources: U.S. Energy Information Administration. National Energy Modeling System, runs refhall.d082611b, ceshallnb.d083011a, refhallhs.d082611b, ceshallnbhs.d083011a, refhallls.d082611b, ceshallnbls.d083011a.

Table B5. Low and high-cost coal scenarios:
The HCES compared to the sensitivity base cases

	2009		2025 Low Cost coal		high Cost coal		2035 Low Cost coal		high Cost coal	
		Base	HCES	Base	HCES	Base	HCES	Base	HCES	
Generation (billion kilowatthours)										
Coal	1,772	2,132	1,095	1,906	978	2,260	878	1,876	447	
Petroleum	41	45	44	46	44	47	45	48	45	
Natural Gas	931	952	1,476	1,071	1,486	1,267	1,783	1,456	2,056	
Nuclear	799	877	933	877	961	874	1,118	874	1,201	
Conventional Hydropower	274	306	323	304	319	314	324	311	319	
Geothermal	15	27	27	25	29	42	48	37	48	
Municipal Waste	18	17	17	17	17	17	17	17	17	
Wood and Other Biomass	38	156	282	180	257	169	268	219	196	
Solar	3	18	19	18	18	21	24	21	26	
Wind	71	156	190	153	301	164	293	158	318	
Other	18	16	16	16	16	16	16	16	16	
Total Generation	3,981	4,703	4,422	4,614	4,426	5,192	4,814	5,035	4,691	
Capacity (gigawatts)										
Coal	317	327	273	308	247	338	279	312	212	
Petroleum	116	86	87	88	89	86	87	88	86	
Natural Gas	351	381	378	383	384	454	444	456	453	
Nuclear	101	110	118	110	122	110	142	110	153	
Conventional Hydropower	78	79	83	78	82	81	83	80	82	

	2009	2025				2035			
		Low Cost coal		high Cost coal		Low Cost coal		high Cost coal	
Geothermal	2	4	4	3	4	5	6	5	6
Municipal Waste	4	4	4	4	4	4	4	4	4
Wood and Other Biomass	7	17	17	17	17	20	20	20	20
Solar	2	11	11	11	11	13	14	13	16
Wind	32	54	64	53	99	57	96	55	105
Other (including pumped storage)	24	25	25	25	25	25	25	25	25
Total	1,033	1,098	1,065	1,083	1,083	1,192	1,201	1,168	1,161
Prices (2009 cents/kWh)									
Credit Price			11.1		8.2		13.2		14.6
Electricity Price	9.8	8.8	10.9	9.4	10.9	9.1	12.1	10.0	13.2
Residential	11.5	10.4	12.5	11.1	12.6	10.6	13.6	11.6	14.7
Commercial	10.1	9.0	11.2	9.6	11.2	9.1	12.2	10.0	13.4
Industrial	6.8	6.0	7.8	6.5	7.8	6.3	9.0	7.1	10.0
Average Delivered Natural Gas Price (2009 dollars/MCF)	7.5	8.0	9.3	8.2	9.1	9.2	9.2	9.4	9.7
Expenditures (billion 2009 dollars except as noted)									
Total Electricity Expenditures	350	365	425	382	424	409	495	434	522
Residential Electricity Expenditures	156	154	176	160	176	172	207	183	218
Household Electricity Expenditures (2009 Dollars/Household)	1,379	1,139	1,304	1,190	1,306	1,171	1,407	1,247	1,487
Natural Gas Expenditures	156	184	242	193	239	225	260	241	290
Electricity Sector Natural Gas Expenditures	34	37	75	43	73	55	79	64	98
Non-Electricity Sector Natural Gas Expenditures	122	147	166	150	166	170	181	177	192
CES Compliance									
Credits Required (percent of sales)			64		64		80		80
Credits Achieved (percent of sales)			63		64		80		78
Generation Achieved (percent of sales)			63		64		80		78
Total Electricity Sales (billion kilowatthours)	3,556	4,148	3,885	4,050	3,881	4,494	4,070	4,324	3,928
Emissions									
Sulfur Dioxide (million metric tons)	6	4	3	4	3	4	2	4	1
Nitrogen Oxide (million metric tons)	2.0	2.0	1.4	2.0	1.2	2.1	1.1	2.0	0.7
Mercury (metric tons)	40.7	30.2	15.6	27.1	12.6	31.0	13.7	24.9	5.9
Carbon Dioxide (million metric tons CO_2)	2160	2417	1501	2220	1430	2583	941	2248	879

Table B5. (Continued)

	2009	Low Cost coal		high Cost coal		Low Cost coal		high Cost coal	
		2025				2035			
		Base	HCES	Base	HCES	Base	HCES	Base	HCES
Macro Economic									
GDP (billion 2005 dollars)	12,881	20,016	19,860	19,973	19,867	25,710	25,591	25,623	25,514
Per Capita GDP (thousand 2005 dollars/person)	42	56	55	56	55	66	66	66	65
Employment, Non-Farm (million)	131	156	156	156	156	171	171	171	170
Employment, Manufacturing (million)	12	16	15	16	15	13	13	13	13

Sources: U.S. Energy Information Administration. National Energy Modeling System, runs refhall.d082611b, ceshallnb.d083011a, refhallhc.d082611b, ceshallnbhc.d083011a, refhalllc.d082611b, ceshallnblc.d083011a.

APPENDIX C. MAP OF NEMS ELECTRICITY MARKET MODULE REGIONS

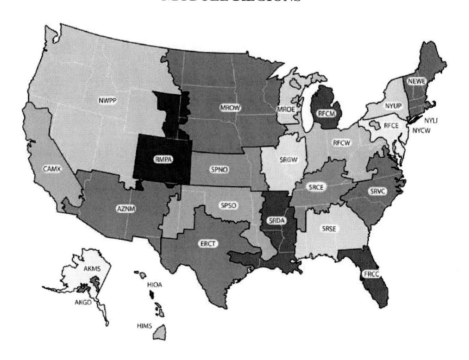

End Notes

[1] The Reference Case in this report includes some revisions to the AEO2011 Reference Case. The primary changes include an improved representation of interregional capacity transfers for reliability pricing and reserve margins. Also, capacity expansion decisions incorporate better foresight of future capital cost trends by including expectations of the commodity price index.

[2] Real GDP and real GDP per capita are reported in 2005 dollars.

[3] The baseline scenarios are: the Reference case, high-cost nuclear, low-cost nuclear, high-cost renewables, low-cost renewables, high-cost gas, low-cost gas, high-cost coal and low-cost coal.

INDEX

D

E